ECOLOGICAL EPISTEMOLOGIES AND SPIRITUALITIES IN BRAZILIAN ECOVILLAGES

This book brings together ethnographic field research on four permacultural ecovillages in Brazil to highlight the importance of spirituality and ecological epistemologies as key analytical tools. It demonstrates that ecological spirituality can, and should, be understood beyond the dichotomy of personal and political, between people and nature, in the field of environmental anthropology.

The book uses a broad philosophical methodology based on the phenomenological theories of Maurice Merleau-Ponty, Tim Ingold and Alfred Schutz combined with post-structuralist conceptions of the relationship between person and world, individual and society. The field research consisted of ethnographic travel, observation and recorded dialogue with individuals based in each ecovillage: Arca Verde, situated in Campos de Cima da Serra; Vrinda Bhumi, a Vaishnava ecovillage in Baependi-MG; Goura Vrindavana, a Vaishnava ecovillage in Paraty-RJ; and Muriqui Assu Ecovillage Project, a secular ecovillage in Niterói-RJ. Throughout the book ethnographic research is woven together with poetic interludes, images, personal narrative experience and phenomenological theory, bringing a new understanding and approach to environmental anthropology as a discipline.

Including a Preface written by Tim Ingold, it will appeal to academics, researchers and upper-level students in phenomenology, environmental philosophy, environmental anthropology, religious studies and social sciences more broadly.

Luz Gonçalves Brito holds a PhD in Social Anthropology from the University Federal do Rio Grande do Sul/UFRGS, Brazil. She is committed to environmental humanities, phenomenology and intersectionality. Her current research interests include political ecology, gender, race and mental health.

Routledge Environmental Anthropology

Environmental Anthropology explores historic and present human–environment interactions, highlighting the link between human-caused environmental problems such as climate change, species extinction, and pollution, with the complex cultural, political, and economic systems that have created them. This series aims to contribute to the growing subfield by providing a comprehensive survey of contemporary topics in environmental anthropology, including food procurement, ethnobiology, spiritual ecology, resilience, nonhuman rights, architectural anthropology, industrialism, and education.

The *Routledge Environmental Anthropology* series welcomes submissions that combine strong academic theory with practical applications, and as such is relevant to a global readership of students, researchers, policy-makers, practitioners, and activists. Please contact Grace Harrison (Grace.Harrison@tandf.co.uk).

Ecological Epistemologies and Spiritualities in Brazilian Ecovillages
In the Labyrinth of an Environmental Anthropology
Luz Gonçalves Brito

For more information about this series, please visit: https://www.routledge.com/Routledge-Environmental-Anthropology/book-series/REA

Ecological Epistemologies and Spiritualities in Brazilian Ecovillages

In the Labyrinth of an Environmental Anthropology

Luz Gonçalves Brito

Routledge
Taylor & Francis Group
LONDON AND NEW YORK

earthscan
from Routledge

First published 2023
by Routledge
4 Park Square, Milton Park, Abingdon, Oxon OX14 4RN

and by Routledge
605 Third Avenue, New York, NY 10158

Routledge is an imprint of the Taylor & Francis Group, an informa business

British Library Cataloguing-in-Publication Data
A catalogue record for this book is available from the British Library

Library of Congress Cataloging-in-Publication Data
Names: Brito, Lucas Gonçalves, author.
Title: Ecological epistemologies and spiritualities in Brazilian ecovillages : in the labyrinth of an environmental anthropology / Luz Gonçalves Brito.
Description: 1st edition. | New York : Routledge, 2023. | Series: Routledge environmental anthropology | Includes bibliographical references and index.
Identifiers: LCCN 2022061632 (print) | LCCN 2022061633 (ebook) | ISBN 9781032458205 (hardback) | ISBN 9781032458212 (paperback) | ISBN 9781003378853 (ebook)
Subjects: LCSH: Ethnology—Fieldwork—Brazil. | Human ecology—Religious aspects. | Ecovillages—Brazil. | Permaculture—Brazil.
Classification: LCC GN346 .B7 2023 (print) | LCC GN346 (ebook) | DDC 305.80072/3—dc23/eng/20230320
LC record available at https://lccn.loc.gov/2022061632
LC ebook record available at https://lccn.loc.gov/2022061633

ISBN: 978-1-032-45820-5 (hbk)
ISBN: 978-1-032-45821-2 (pbk)
ISBN: 978-1-003-37885-3 (ebk)

DOI: 10.4324/9781003378853

Typeset in Times New Roman
by codeMantra

For Keron Ravach

Contents

Illustrations

Acknowledgements

I could thank many people and situations, and I would still forget someone. Professors, folks, unknown people and rivers. I know who you are, and I am deeply grateful.

I thank you, Mother, for working hard to guarantee a good education for me and my sister.

Thank you, Sara Gonçalves Brito, for being the reflection of the strong and whole woman I became.

Thank you, Juliana Martins Pereira, who first told me of the word Anthropology. In 2013, you asked me to share my breath with the world. And here I am.

Thank you, Thaís Nascimento, great musician, for the honour of making me your musical partner.

I thank Gustavo Chiesa, my intellectual partner, with whom I share rare dialogue and ontological affinity.

Many thanks to all interlocutors I met during my ethnographic travels. I could never completely express with concise words the breadth and profoundness of your contributions. For this reason, the notes of my melodies and the verses of my poems will always testify the generosity of the Earth.

I express my gratitude to my adviser, Dr. Ari Pedro Oro, who has supported me and helped me like a nurturing father.

Thank you very much, Dr. Tim Ingold. Our dialogue is an honour for me!

Many thanks to Dr. Isabel Carvalho, Dr. Eduardo Dullo and Dr. Rodrigo Toniol for your precious contributions to my work.

I thank all the people who welcomed me in California, particularly Evo Wang and Debbie Kinsinger. Many thanks to Jennifer Freeman and her group of Gaia activists, who understood the importance of this book.

Thank you, my friend Tree of the Memory Lake (Porto Alegre).

Special thanks to Heloá, Sofia and David, the three children who talked to me whilst I was writing these lines at the Ecological Park Tietê (São Paulo).

Thank you, Marcus Vinicius, for being exactly who you were.

I am grateful for the opportunity of being alive, which nourishes my consciousness through the lifelines with which I am entangled.

Foreword

Considering the present condition of the Earth, of this planet we call home, it is hard not to experience an acute sense of despair. Its eight billion inhabitants remain in the grip of a system of production, distribution and consumption that is relentlessly laying waste to its once bountiful lands and oceans, leading to the grotesque enrichment of a very few while immiserating the great majority and leaving them impoverished. Its nations are divided into states whose politics reward stupidity and brutality with powers of oppression that are determined to crush every green shoot of renewal. A widespread feeling of impending catastrophe obscures the fact that for many, the catastrophe has already occurred, as their lands have been flooded, homes bombed and forests burned. These are indeed terrifying times. Nor is our terror assuaged by the assurances of historians that war, plague, famine and environmental catastrophe have been an ever-present backdrop to human societies for as long as anyone can recall. And yet here we are – more of us than ever before, living to a greater age, on average, than ever before – but for how long will it continue? Would we even want it to continue?

It is easy to feel helpless. It is beyond the capacities any one of us to change the world, and even collective action has its limits. But what if we were to change our *relation* to the world? Might part of the problem lie in our tendency to see the world as something outside ourselves, as an externality so gigantic as to be immoveable by any of its Lilliputian inhabitants? The flipside of this tendency is a kind of locked-in syndrome, which causes us to see ourselves as the prisoners of our bodies, with our minds inside our heads, deprived of immediate access to what lies beyond. To make up for this deprivation we clutch at straws, clinging with our fingertips to whatever opportunities the outside world affords. Yet there are those who would insist that the prison is a figment of our imagination, conditioned by an authoritarian system of education that seeks to mould every individual to the shape of society. To change society may be a tall order, but we surely have the power to change ourselves. This is the power of spirituality, and it is the theme of this engrossing book.

One of the more debilitating aspects of contemporary global capitalism lies in its drive to convert those qualities that confer intrinsic goodness on communities of the living – qualities of grace, creation and spirit – into marketable commodities to be consumed in individual isolation. Grace becomes gratification, creation creativity and spirit spirituality. For many people today, the very idea of spirituality conveys the self-satisfaction of an affluent middle class, with its penchant for mindfulness, aromatic fragrances, well-being therapies and self-help workshops. Luz Brito, however, is on a mission to unfasten the idea from its lifestyle associations with the New Age, to restore the *spirit* to spirituality, and to find in it a necessary foundation for hope. Spirit, properly speaking, is the creative force of a world that is continually coming into being, or worlding, in and through the productive endeavours of its inhabitants. A life that is enspirited is not lived in isolation, or expended in the immediate gratification of its own desires. It may be solitary, but solitude is not isolation. For even as the solitary mind admits the world's fullness into its deliberations, the body mingles with its worlding. And in life, mind and body are one.

No more than grace and creation, then, is the spiritual a property of individuals. It is rather intrinsically ecological. This is not, of course, the ecology of ecological science, which continues to treat the world as its object. For a spiritual ecology the world is its *milieu*, an unfolding field of relations and processes within which every creature, whether human or non-human, comes into being and lives its life, contributing through its actions to the lives of others. By tuning in to these relations and processes, and cultivating our powers of attention, spirituality merges with ecological thinking in the perception of an environment that is immanently *sacred*, immediately erasing any distinction between interior self and exterior world. To transform the self is therefore, also, to transform the world. But there's also a challenge in this. For self-transformation leaves unscathed the forces of oppression that continue to yoke the majority to a cycle of environmental destruction, driven by global capitalism, from which there seems no escape. Since these forces are ultimately political, only by concerted political means can they be overcome. How can spirituality be given a political edge, so that it can cut through into wider societal transformation?

That's the challenge of this book, in which Brito travels to four "ecovillages" in Brazil, all of them experiments in alternative lifeways grounded in principles of permaculture. As a volunteer in each village, she embarks on a series of dialogues with her new-found companions, founded in a deep respect for what she has to learn from them. This doesn't, of course, mean accepting everything they say. For her aim is not so much to interpret or explain their ideas for us, her readers, as to join *with* them in a conversation from which mutual understanding can potentially grow. This calls for a stance that is not only humble but also relentlessly questioning and critical. In many cases,

Brito's conversation partners would disagree among themselves. And if they are entitled to disagree with one another, so too is she with them. A few, at least, have apparently managed to wrap themselves up in a somewhat self-serving comfort blanket of ever-ramifying delusion. At what point, then, should we cease taking them as seriously as they seem to take themselves? If we fail to do so, do we not run the risk of deluding ourselves as well? Read on, and decide for yourself!

Tim Ingold
University of Aberdeen

Prelude

Spirituality beyond the New Age

Whenever spirituality is understood only as a private and individual experience, researchers misunderstand the different facets of this relevant category of modernity. The idea that spirituality concerns this human dimension circumscribed to self-realisation per se or well-being per se seems incomplete. That is one of my research problems, which is unfolded in different questions.

How does one find a wider concept of spirituality beyond the dichotomy between the personal and the social? Which are the features of spirituality as a historically situated category? What are the identifiable processes by which spirituality became a socially and anthropologically relevant concept? Which are the possible different notions of spirituality?

The axial argument developed alongside this dissertation is that spirituality is not opposed to materiality. The conceptualisations of these people with whom I have talked, as well as the philosophical systems in which their conceptions are embedded, indicate (and maybe also demonstrate) that there are other possibilities. One such possibility is the conception of spirituality as an experience of transformation of the person (and the world) in the immanent lived world. That is, nevertheless, one amongst other possible different conceptions, some of which will be observed within this dissertation.

During my bibliographic research, I noticed that spirituality, in general, is a very important category for the studies that associate spirituality with the New Age movement and with health in a broad sense. A fundamental reference in this set of studies is the in-depth work of Toniol (2015a, 2015b, 2017).

My first contact with the notion of "New Age" happened while I was doing fieldwork for my master's research. I had attended a seminar supported by Centro Espiritualista de Umbanda Pai Joaquim de Angola, situated in the city of Goiânia-GO. The seminar, which took place in 2014, focused on a single main theme: planetary transition. Many spiritual leaders joined the event. My subject of research was different and, therefore, I did not approach the topic.[1] However, some dialogues with those people in the fieldwork revealed the term planetary transition as a universal, periodic process of change in the state of

DOI: 10.4324/9781003378853-1

human consciousness. The people who were umbandistas, followers of this specific religion Umbanda, told me that changes in human consciousness were related to geophysical, social and political transformations. The idea of planetary transition meant a specific convergence of ecological thought and the New Age.

When I read the articles of De La Torre (2011, 2016), I understood the historicity of the utopian idea of planetary change. According to this New Age idea, the planet is passing through an unprecedented cyclical change. The physical and chemical process was understood as a response to the people's current state of consciousness. Even climate change would be related to the egocentric Euro-American way of life and understood as a warning of the responsibility of self-transformation, so the planet would respond back to human's attitudes, generating a softer environment.

De La Torre (2011, 2016) emphasised the relevance of the book written by Marilyn Ferguson, *The Aquarian Conspiracy*, which was first published in 1980.[2] The wide reception of the book catalysed and spread the idea of the Age of Aquarius beyond the circles of counterculture. The book manifested the sensibility of the times, the *Zeitgeist* of the second half of the twentieth century, announcing the importance of the personal and inner change *pari passu* with the social action leading to collective change. Ferguson's book registered a pervasive historical and social phenomenon:

> Many of the activists were turning inward, a direction that seemed heretical to their comrades in the conventional Left. They were saying that they could not change society until they changed themselves.
>
> (Ferguson 1980: 58)

However, if we are to understand the New Age phenomenon,[3] it is possible to trace back another line, even earlier. Seeds of the New Age movement were planted by the end of the fifties, when the generation of North American young beatniks, leaving their hitchhiking trips full of jazz and blues, gave way to the generation that searched for magical experiences with mushrooms and other psychoactive substances. During that period, global political and social transformations were happening. The experience of the Second World War was disastrous, entailing processes of cultural deterritorialisation, trends which continued during the equally disastrous Vietnam War. Besides globalisation, the world became divided in three – the first-world countries, fully developing their expansionist capitalist market systems; the communist second world, raising the flags of collectivity and working-class sovereignty; and the third-world countries, the rest of the developing countries that provided much of the raw material and natural resources (fundamental to the maintenance of the richness of the first world).[4]

The New Age movement, often gravitating towards alternative communities, emerged as a countercultural critique of the capitalist values of accumulation. It recovered Eastern, shamanic and neo-pagan practices that valued the connection with Earth as a feminine spirituality (Carozzi 1999a). The people who shared the same critical views, rose, in the cosmopolitan societies, as "pollinators of a global consciousness and of a universal consciousness" (De La Torre 2016).

The New Age movement – even if it later became associated with certain liberal logic of a bourgeois ethics propagated by the schooled middle class (D'Andrea 2000: 41) – allowed for non-institutionalised spiritualities, by which each person felt free to produce particular bricolages and synthesis of symbols, concepts and techniques, according to their trajectories (Camurça 2003). On the level of the experience of the sacred, the New Age disseminated the individual autonomy to choose practices borrowed from different philosophical systems and plural religious traditions (Carozzi 1999b). The New Age has been considered a specific way of dealing with the sacred that refused substantial and fixed identities, inextricably reflecting the nomadism in the globalised world (Amaral 1999, 2000).[5] The New Age implies scales, processes, people, collectives, agendas, concepts, practices and techniques that try to respond to the crises of the contemporary world – ecological, existential, institutional, political, cultural, social and epistemological crises.

An attentive reading of the bibliography concerning spirituality and alternative communities clearly reveals the aprioristic association of spirituality and the New Age.[6] The problem with this assumption is the essentialisation of spirituality. Methodologically, that assumption obnubilates the fact that spirituality is a multiple phenomenon whose conceptions are not even nearly homogenous.[7]

If spirituality is entirely scattered in the world, where is it possible to identify it, unless in concrete and specific contexts? This precaution prevents losing the description capacity of the category. The category of New Age, for example, has been applied to numerous phenomena, including the extension of the category to embrace the process of incorporation of New Age features by religious and secular institutions – a process which was called "renewalisation" (Toniol 2015a). This kind of hyperbolic conceptual enlargement of the analytic category is an example of the situation where "the multiplicity of observed realities was imprisoned by the fiction created to describe it" (Toniol 2015b: 20).

Even the understanding of New Age as an adjective and not as a noun (Amaral 2000) has not been sufficient to avoid the exhaustion of the category. Although the category New Age has worked as a methodological key in social research at the turn of the century, there was a quantitative decrease of texts regarding the topic ten years later (Guerriero, Stern and Bessa 2016). In fact, the category of New Age became almost ubiquitous by the diffusion of New Age ideas and techniques. And that ubiquity can practically become an analytical barrier, even if there is no evidence that the propagation of the New Age *ethos* was the only factor generating the reduction in the research.

Unsurprisingly, the notion of spirituality has been easily connected to counterculture discourses, which were in general very close to what has been understood as New Age. More than 20 years ago, Hanegraaf noticed that

> The term "New Age" has entered the standard vocabulary in discussions about ideas and practices regarded as alternative *vis à vis* dominant cultural trends, especially if these ideas and practices seem to be concerned with "spirituality".
>
> (Hanegraaf 1996: 1)

Thus, the so-called "New Age way of dealing with the sacred" (Amaral 1999) – highly focused on the interiority and human potential development – has also been interpreted as an expression of post-modern psychologisation, reflexivity and globalisation (D'Andrea 2000) and also of the "sacralisation of individual autonomy" (Carozzi 1999a).

The tight relation between the New Age phenomenon and the notion of spirituality can be clearly seen in the conceptions of the New Age, especially those according to which the main characteristic of spirituality is the same of the New Age: the "strictly personal" relation with the sacred (De La Torre 2011).

That conception is, in some respects, shared by Heelas (2006), according to whom the growth of the spiritualities of inner life (or New Age spiritualities) coincided with the decline of certain religious traditions, in the context of modernity. The validation of the modern individualist ethics – brought to paroxysm – was unleashed by the development of the autonomous self and the subjective turn of modernity.

Evidently, I am not affirming the impossibility of analytical association between New Age and spirituality. Both notions are, for example, intrinsically linked to the countercultural and autonomous experiences lived in alternative communities (De La Torre 2016). The conjunction of the Human Potential Movement in Esalen (California) and the theosophical principle of One Truth in Findhorn (Scotland) – from which emerged New Age notions (Carozzi 1999b) – demonstrates that the so-called intentional and/or alternative communities are fecund places where both New Age and spirituality can be understood. I am not interested in fixing the "true origin" of the phenomena, but rather in recognising the relations between the notions of spirituality and New Age and dissociating them, at the same time. Such a procedure, I suppose, renews the potentiality of spirituality as a category, unfastening its seemingly inexorable ties with the New Age. Hence, I propose emphasising the concepts and experiences lived by my interlocutors in these "communities for the difficult times" (Hervieu-Léger and Hervieu 1983) as the soil in which different notions of spirituality grow.

The ecological in the ecological epistemologies

But, at this point, the reader may be questioning: what is the relation between all of those issues and ecology? And this is the answer: ecological epistemology and spirituality can be considered as two flows of the same river. Let me expose my hypothesis in a less enigmatic way.

Notice that I am not referring to ecology, but to ecological epistemology, the preferred term for this study. Ecological epistemologies concern:

> A region of the contemporary philosophical and theoretical debate that involves authors whose disciplines and theoretical options vary. Those authors share a common point, which is the effort to overcome the modern dualities, such as nature and culture, subject and society, body and mind, artifice and nature, subject and object.
>
> (Steil and Carvalho 2014: 164)

Ecology as an academic discipline consists of the study of the interrelations between organisms that set systems. The prefix "eco-" derives from the Greek word *oikos*, and means "house", "home" or "place where one lives". The suffix "logy" also derives from a Greek word, *logos*, which stands for "study". Etymologically, the definition of ecology is "the study of the organism 'in their homes'" (Odum 2001: 4). Moreover, ecology observes and describes the patterns of the relations between organisms and their environments.

> Ecology is usually defined as the study of the relations of organisms and groups of organisms with their environment, or the science of the interrelations that connect the living organisms to their environment.
>
> (ibid.)

Ecology is a systemic science, whose language is concerned with wholes. Nevertheless, its methods and techniques are not the most important topic here, but rather its attention to "the pattern that connects" living organisms (Bateson 1979). Such a premise is not an exclusive concern of ecology as an academic discipline because it is disseminated to other sciences and beyond academia, especially in Euro-American societies. Ecological epistemologies underlie different discourses – academic or not – on the interrelation between living organisms. Methodologically, that understanding enables the localisation of ecological epistemologies under the surface of my interlocutors' discourses.

The aforementioned premise was not the only notion that has been spread. On the other hand, there is also the speculist assumption that "has given us the imagery of the world as a globe" (Ingold 2002: 212). The idea that the world and its ecosystems could be perceived "from the outside" and "above" is partially derived from the astronomical technological instruments. Such a distant perspective later enabled the population to imagine the Earth from a paradoxically reductive and macroscopic perspective.

There are at least three different conceptualisations regarding how humans perceive the world. The first mode of perception of the Earth as a globe exteriorises people. It refers to the "anthropocircumferentialism" (Ingold 2002) through which human life is detached from its participation in the environment. From the moment that people sense the Earth where they inhabit is outside themselves, the result is a self-perception as an entity separate from the world. This angle allows for the disjunction between humans and the environment, whilst nature is maintained inside of the world and society is set aside of the world, as if society were upon nature. On the other hand, the concept of Earth as a globe is also related to the anthropocentric disjunction of the humanity that supposedly would live on a plane distinct from the sphere in which the other beings live.

The second form of perception of the Earth is representational and isomorphic. This Cartesian notion of perception is grounded on the reduction of the world by the individual mind. In this case, the perception would involve the internal representation of the image of the world in the brain. The extension of the perceived world is reflected on the eyes through luminous rays that strike the retina. Then, the optic nerves capture and reconduct these rays – which Descartes (1985) called "animal spirits" – to the pineal gland in the centre of the brain in the form of inverted images. The representation[8] of the reduced object inside the brain is understood as having the same character of the perceived object, which is situated outside the perceiver.

The third notion of the perception process is derived from Kant. Besides beings detached from the world, the subject who perceives cannot apprehend the world-in-itself, which is incognoscible. The categories of the understanding that the subject projects upon the world do not correspond to the characteristics of the world and, as such, everything that seems possible to be known is the phenomenon happening "at the meeting point between the inaccessible things in themselves and the categorizing work" (Latour 1999: 72). The known world, in this case, would be just a superficial reflection of an idea.

The most interesting notion of perception that I know is a different one, which Ingold (2002) has cultivated. According to this notion, the perception of the world does not take place from the outside – or from above, as if the person were separate from the environment, looking at the world – but rather through one's engagement with a world inhabited from within. This modality of perception of the environment corresponds to the concept of the environment as a lifeworld (see Ingold 2002: 209). Notwithstanding the foundations that the other notions can provide to ecology, this "ecological approach" described by Ingold is the ground to my understanding of ecological epistemologies. The immersion of the person in the environment can be metaphorised as the little grain of sand on the shore. The grain, despite being different from the sand, is in the sand as the sand is in the grain. As well as the sand is on the shore and the shore is on the sand.

Finally, ecological thought and spiritualities are different ways of perceiving and searching for the connections between things and beings in the world. Both can be conjoined by different conceptions on the relations between people and the world. More broadly, ecological epistemologies and spirituality express the search for connection between parts and wholes – such as organism-system, person-society, human-nature, person-deity, individual consciousness-Cosmic Consciousness or individual soul-Universal soul and so forth.

My fieldwork, ethnographic notes and interviews/dialogues demonstrate that there are different notions of spirituality intricately linked to ecological epistemologies. I found three notions of spirituality, repeated in different contexts – spirituality as the search for the inner transformation that resonates in the world as social transformation; spirituality as the search for connection with the world; and spirituality as the cultivation of a sensitivity to the environment.

Epistemic option and phenomenological option

The different endogenous conceptions explored along this dissertation are inherently part of systems of knowledge: epistemologies. My attention to the endogenous conceptions amounts to bringing them to the anthropological formulation. Although the endogenous conceptions present their own analytical level, they work differently from anthropological knowledge. These endogenous conceptions – i.e., conceptions embedded in the systems of knowledge of the collectives and people with whom I have researched – are not considered as mere data to be analysed, but rather as concepts to be described; conceptions from which anthropologists can learn. This process of juxtaposition of endogenous concepts and analytical concepts is not a hylemorphic procedure where the anthropologist explains other people's ideas, as if those ideas were the material that fills the anthropological conceptualisation or the substance moulded by the anthropologists by means of anthropological theory (see Viveiros de Castro 2013: 475). On the contrary, I recognise that the conceptions presented by the people with whom I dialogue are embedded in an epistemology. Such a shared system of ideas and practices regards the acquired knowledge on the existent things and beings, and people use it to guide their action in the world.

My position, in a certain way, coincides with the differentiation of culture and "culture":

> I firmly believe in the existence of internalised schemes that organise the perception and the action of people, guaranteeing some degree of communication in social groups. Something like the so-called culture. But I also believe that this culture does not coincide with "culture". There are significant disparities between them. This does not mean that their content is different, but rather that they do not belong to the same universe of discourse.
> (Carneiro da Cunha 2009: 313)

I agree that the anthropological discourse on culture receives quotation marks when it becomes an analytical concept for the discipline. Evidently, there are two distinct universes, the anthropological and the endogenous. Nevertheless, I tend not to previously deny that my interlocutors do not have a "reflexive meta-discourse on their own culture" (Carneiro da Cunha 2009: 373). They do express, more or less explicitly, knowledge regarding the ideas and practices they share with other people.

Nonetheless, I understand that the so-called culture is a result of the dialectic of engagement and disengagement of the person in the world. This dialectic of engagement and disengagement engenders the reflexive knowledge that anthropologists tend to understand as culture. Whereas our interlocutors have neither the time nor the inclination to follow this movement, anthropologists learn to flow with it by means of the engagement in the field and the disengagement of the distanced reflection. Seen in these terms, culture is not the dichotomic pair of nature.

> The separation of nature and culture, as domains respectively of matter and mind that humans in their activities must perforce seek to bridge, far from existing *ab initio,* is a *consequence* of disengagement, of the turning of attention, in thought, reflexively inwards on the self rather than outwards on the world. Now although humans are undoubtedly capable of adopting such a contemplative stance from time to time, no one – not even a monk or a philosopher – can permanently live like that.
>
> (Ingold 2005: 96)

The person's reflexivity regarding their own ideas and practices involves a certain disengagement from their acts or thoughts. An example? While I was doing fieldwork in the ecovillage Arca Verde (São Francisco de Paula-RS), I was also learning some planting techniques for an agroforestry system. I had asked my instructor to tell me how to identify the best moment to plant. He looked at me, with his stuck hands above the garden plot and a seedling in his fingers. Then, he started to explain the influence of the moon on the flows of the water through the soil, through that seedling and through the plants in general. I observed that he literally "stopped to think". And that observation is not banal at all. The technique he had been using was evidently rooted in a certain knowledge, but his reflection regarding the knowledge and the technique was possible because he interrupted his engagement with the earth. The knowledge that reflects on itself rises exactly from the suspension of the natural attitude with which humans in general engage in their life-worlds, even if it is possible to contrast "the world of scientific thought with the life-world of natural experience" (Schutz and Luckmann 1973: 21). Scientific knowledge as well as everyday knowledge necessarily rise from the gesture of stopping-to-think. Nevertheless, such a gesture is, paradoxically, a movement in which the perceiver enacts a deeper engagement with the living world.

For me, knowledge became a category of the greatest importance because I understand *culture as lived knowledge*. Culture is the knowledge of the life movement, and culture is born from the movement of life and the temporary pause in the movement of life. Culture cannot only be understood as the set of simple beliefs or representations constructed by people in order to deal with their worlds, make sense of their worlds or guide their modes of action in those worlds. Consequently, cultures – because it is necessary to recognise their plural character – do not regard particular worldviews set upon a shared nature. The notion of culture as lived knowledge implies the perceptual, processual, transformative, lived, phenomenological character of human engagements with the world.

I am not exactly postulating a new concept of culture, but rather presenting a heuristic notion for this study. In fact, I arrived at this understanding after dedicating some years of my attention to the different cultural epistemologies underneath the discourses expressed by the people with whom I dialogue.

Knowing implies inhabiting and living the world. People, instead of knowing the world in which they will inhabit or constructing representations of the world in which they will live later, inhabit the world and know it. And even if those abstract representations are really constructed, they would never exist prior to the experience. Humans do not set ideas down on the world before the experience of the world (Ingold 2005), but they rather live the experience in the world and construct ideas regarding life, whilst they are knowing the world. If the world is revealed through the lived experience, life happens and transforms itself in the world, and people know about life in the world. "It is through such attentive engagement, entailed in the very process of dwelling, that the world is progressively revealed to the knowledge-seeker" (Ingold 2002: 216). Thus, one's ideas on the world concern the life experienced in the world; in a way, the ideas of the person are the very person.

Furthermore, if the idea is, as Bateson (1972) suggested, "a difference which make a difference", then information (and I would add knowledge) is inherent to the processes of perceiving and thinking because the knowledge is organised as a relation of alterity and differentiation of the aspects perceived in the world. If the knowledge is contained by the mind, it is also contained in the world because the mind is not separate from the world.[9] During the process of perceiving and thinking – knowing – the person creates a relation of alterity with the world. The person becomes a perceiver of the world: the world is perceived not because it is outside, but rather because knowing engenders the knower as well as the known. Person and world are interrelated. Hence, each difference in the world perceived by the person in the world is a difference perceived in their own mind and in the world's mind. The mind is not separate from the world because while perceiving the world, the mind perceives itself.

Besides an epistemic theoretical option, this study as a whole presents a phenomenological option.

Methodologically, the epistemic option is the description of the different conceptions on the way people relate to the world, in different places. It is also the description of how the world responds to such relations. Regarding the phenomenological option, the point is understanding how people engage in the world according to the different conceptions that influence their modes of action.[10]

There is another wider theoretical implication, which is rejecting the reduction of the anthropological procedure as a search for external social facts, as if anthropologists could settle for the notion of phenomena outside people, like if phenomena were incognoscible things in themselves or a reality independent of the ideas we have about them. In other respects, the phenomenological option avoids the belief that the researcher's ideas correspond completely to the observed phenomenon.

Thus, it is necessary to humbly accept that the anthropological reflection, even though complex in its formulations and wide in its partial connections, is always linked to situated points of view of the researchers, who perceive the things and beings in the world, and who also perceive themselves in the world, from a specific place. Ergo, I want to emphasise that my reflexive usage of different analytical languages alongside this text is inherently an issue to me. I have worked with different analytical concepts without agreeing completely with them or adhering to any specific theoretical framework. Thereby, I distance myself from academic sectarianism, which, to my eyes, hinders the thought. In this regard, I understand that "the method must fit to the object and the object must not obey the injunctions of the method" (Bastide 2007: 140).

How the field was chosen

The methodological path was delineated through the bibliographic research regarding the theme of the New Age.

The studies of Amaral (1999, 2000) and Magnani (1999a, 1999b, 2000) were fundamental to my localisation of the subject of this dissertation. Through these studies, I could also situate the places where I would find distinct conceptions of spirituality. The New Age movement is diffuse, and so is the spirituality associated with it.

The empirical universe of Amaral (2000) encompassed workshops and the Alternative Communities Meeting (ENCA). From those empirical happenings, Amaral (1999) conceived the notion of New Age as a field of very diverse discourses, which included the proposals inherited from counterculture by alternative communities, the alternative self-development therapies, the curiosity for the occult, the ecological discourses of the cosmic subject's encounter with the inner essence, the sacralisation of nature and the marked-oriented reinterpretation of "inner perfection" (offered to employees at capitalist bureaux).

On the other hand, Magnani (1999a, 1999b, 2000) was interested in the "webs of sociability" of "neo-esoteric circuits" in the cities. This author registered the set of spaces where many holistic practices were offered. Such practices targeted the emotional, mental and physical levels of people (i.e., spirituality as cultivation of self). According to Magnani, there were five types of those spaces: (a) esoteric groups; (b) integrative centres; (c) specialised centres; (d) individualised spaces; and (e) stores. Amaral (2000) would also add alternative communities and workshops.

The crossing of all those places and discourses rendered a methodologically useful sheet.

Places	*Propagated discourses*				
	Self-development therapies	*Curiosity of the occult*	*Ecological discourse*	*Sacralisation of nature*	*Motivational speech in offices*
Esoteric groups	X	X	X	–	–
Integrative centres	X	X	X	X	–
Specialised centres	X	X	–	–	–
Individualised spaces	X	X	X	X	–
Stores	X	X	–	X	–
Workshops/ meetings	X	X	X	X	X
Alternative communities	X	–	X	X	–

Every line stands for the places and every column stands for the discourses. The places in which each discourse is propagated have been marked. Nevertheless, it does not mean that, for example, specialised centres cannot offer motivational speeches for capitalist offices. Alternative communities can support self-development therapies, too.

For my research, the relevant equation refers to places that present ecological discourses and notions of sacralisation of nature. Thus, it results in integrated centres, individualised spaces, workshops, meetings and alternative communities as the privileged empirical contexts where to find concepts on spirituality and ecological thought. However, this dissertation is rooted in the fieldwork in alternative communities of a specific kind, the permacultural ecovillages.

The exploratory phase of the research involved not only the reading of bibliography concerning the New Age, but also the search for spiritualist text regarding the planetary transition. Among them, there was the notion of "end of the world", banally rumoured in websites. Sensationalist news and articles presented specific and alienating appropriations of the scientific discoveries. I observed that apocalyptic discourses were spread on the contemporary imaginary, often translated in frightening language. The idea of the end of the world regards the diffuse fear in Euro-American societies in the face of the possibility of spatial-temporal dissolution (Danowski and Viveiros de Castro 2014). I then discovered that such discourses, deeply dystopian and depoliticised, engendered the sense of generalised social demobilisation that some authors call "ecologies of fear" (Swyngedouw 2011). That thread of discursive formations on cyberspace, despite its relevant effects over the minds, was left aside.

By that time, according to the bibliography I had read and the people with whom I had had a dialogue, I circumscribed the term spirituality to the intimate experience of religion in its broad sense of *religare* (reconnect). But I would not have seen clearly how spirituality was related to ecology without leaving aside the notion of climate change, which seemed to obscure the comprehension of the phenomena that arose from the field. It is important to note that the New Age – now understood also as the ethics embedded in holistic values, notions of immanent sacredness and centrality of the self (Guerriero, Stern and Bessa 2016) – has remained as the thematic background of this dissertation. Amidst that transformative process of the concepts, I decided – maybe an adequate choice – to dialogue with people who live "spiritual" experiences and, instead of closing their eyes in the face of the capitalist system turbulence, search to act in the world by means of environmental activism. Thus, I was able to verify empirically that spirituality did not concern only the cult of the individual, the body and the well-being (Luiz 2013). The people with whom I met were not confined to their apartments, with eyes closed, sitting in lotus position in their living rooms.

The two-week fieldwork experience in the ecovillage Green Arc (Portuguese, "Arca Verde") was fundamental to mature the research (Chapter 1). Ecovillages are places in which the flows of spirituality and ecological epistemologies debouch. I discovered that ecovillages would be excellent places where I would talk to people and understand their experiences with the environment and the Earth. I would also understand how those experiences influence people's perceptions of the environment and the world. The very experience of living in ecovillages would be considered a political position and, even if the reasons why the choice of living there were not intrinsically related to ecology, the argument that spirituality regards much more than only the autonomous self would be as clear as the sunny sky on a springish day.

In 2018, between July and August, I spent 40 days ethnographically travelling. I became a volunteer again, in three ecovillages, recording my dialogues with 20 people. The first place was *Vrinda Bhumi*, a vaishnava ecovillage in Baependi-MG (Chapter 4); *Goura Vrindavana*, another vaishnava ecovillage in Paraty-RJ (Chapter 3); and, in Niterói-RJ, a grange called Muriqui Assu Ecovillage Project (Chapter 2). While the vaishnava ecovillages were philosophically and religiously oriented, Muriqui Assu was secular.

Vrinda Bhumi was discovered while I was acquiring some organic produce for the week, in the second semester of 2017. Some people with books and

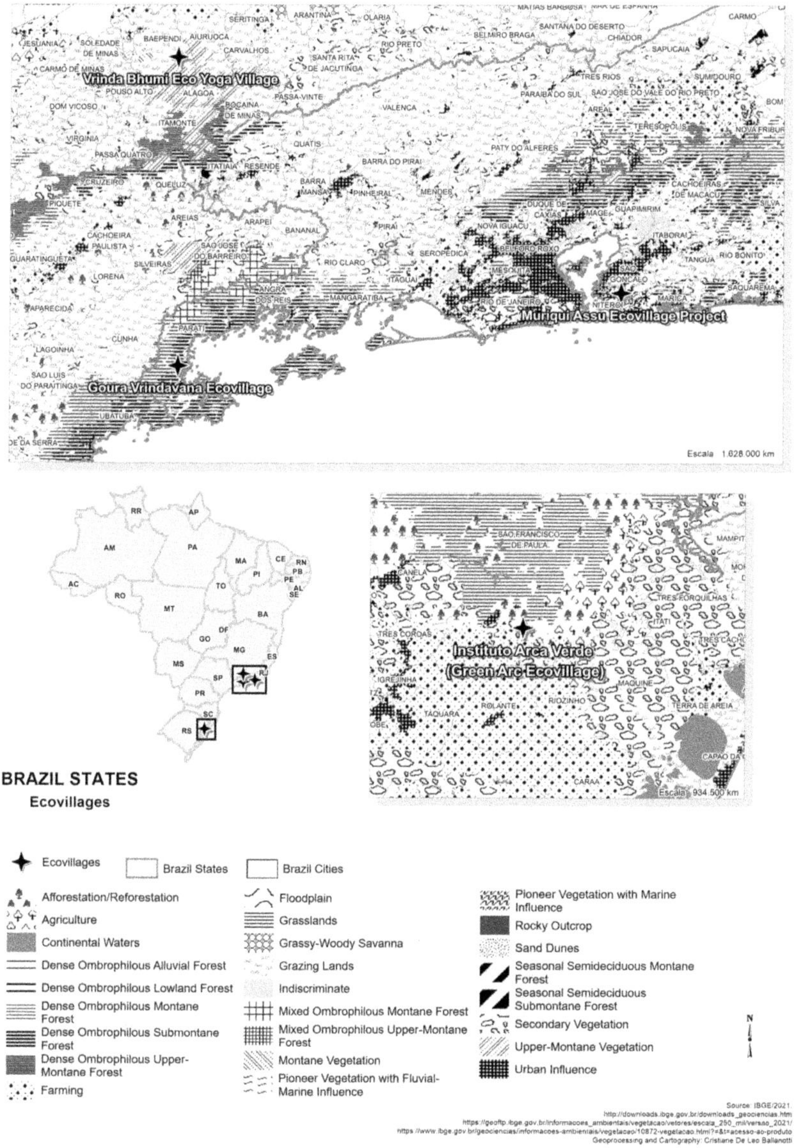

Figure Prelude 1. A piece of Brazil

brochures in their hands, in peculiar attire, approached me. I talked to this young woman who had been visiting the ecovillage recently. I told her I was a researcher and, briefly, described my subject to her. Later, I looked for more information on the internet. Each time I contacted each ecovillage, I was very clear about my research intentions during my stay. The other ecovillages were also discovered on the internet and the keywords I used as a filter were "ecology", "ecological", "spiritual" and "spirituality". The website of Muriqui Assu did not mention the spiritual orientation of the ecovillage. Nevertheless, I would argue that spirituality was just not explicit there.

Notes

1 My master thesis regards the umbandista knowledge constructed in religious temples of Umbanda around the country. In the book that resulted from the research, I presented the epistemic approach to religion (Brito 2019).

2 De La Torre provided a precise description of what turned out to be the New Age:

> The New Age proclaims the end of a rational and material era guided by the Pisces constellation. This will also be the end of the Christian era which, after two thousand years of planetary influence, will give way to the new age of Aquarius, marked by creativity, mystic experience and harmony between humans and nature. (2011 :3)

3 D'Andrea (2000) understood the New Age as a form of contemporary spirituality without churches, leaders and frontiers.

4 Some anthropological perspectives on globalisation can be found in Featherstone (1990), Inda and Rosaldo (2008) and Csordas (2009).

5 In a global context, cultural exchange can be confused with cultural appropriation and mercantilisation of the sacred. According to Steil (2001), the globalisation process generates the interculturality between distant symbolic systems, but also the mercantilisation in the religious field, i.e., the market-oriented vision of the religious field to the detriment of its sacred dimension.

6 Some studies explore the methodological task of tracing genealogies of the category of spirituality in different national contexts (Veer 2009; Toniol 2015b; Viotti 2018; and Gonçalves Brito 2020).

7 Another problem in the research of spirituality is the analytical association with the notion of religion. Consequently, the same problems found by researchers in the universal concepts of religion are present in the concepts of spirituality. According to Bender and McRoberts (2012), "social scientists frequently juxtapose spirituality to religion and identify the former by way of what it lacks in comparison to the latter" and, therefore, spirituality "would appear to lack institutions, authority structures, community, and even history" (2). Whenever researchers identify spirituality in relation to the set of assumptions in the category of religion, they reiterate the Euro-American model of religion, which has been proved to be limited (Asad 1993; Meyer 2014).

8 The infamous "critique of representation" in post-modern anthropology (Clifford and Marcus 1986) involves a strong refusal of the Cartesian dichotomy subject/object. Although the term has been presenting epistemological implications, the common reading of such a debate tends to understand it only in its political sense (Who can represent? Who/What can be represented?). Apropos, the post-modern cultural critique, if solidly understood, enables us to reach poliphony and dialogue in ethnographic texts.

9 I am not referring to the concept of cognitive map, according to which there is "an isomorphism between structures in the world and structures in the mind" (Ingold 2002: 220). My point is affirming the phenomenological continuity between world and mind. The mind is neither circumscribed to the brain nor limited by it. On the contrary, the mind is projected all over the body, towards the environment.

10 The epistemic option involves the language of post-structuralism, whose founding concept is "relation". The phenomenological option deals with "engagement". On the first level of analysis, the premise is the possibility of description of conceptions on the relation between people and the world, nature and society, mind and nature, and so forth. On the second level, the premise is understanding how the person knows those relations through the engagement with the world. If people know the world through their engagement in this world, then it would be possible to assert that such an engagement does not exclude the relations with the environment, which are created in the act of knowing. Therefore, the perception of the environment can be understood as engagement as well as relation because the perception regards "the unfolding of a field of relations established through the immersion of the actor-perceiver within a given environmental context" (Ingold 2002: 220).

References

Amaral, L. 1999. "Sincretismo em movimento – O estilo Nova Era de lidar com o sagrado". In *A Nova Era no Mercosul*, ed. M. J. Carozzi. Rio de Janeiro: Vozes, pp. 47–79.

——— 2000. *Carnaval da Alma: comunidade, essência e sincretismo na Nova Era.* Rio de Janeiro: Vozes.

Asad, T. 1993. *Genealogies of religion: Discipline and reasons of power in Christianity and Islam.* Baltimore e Londres: The Johns Hopkins University Press, pp. 27–54.

Bastide, R. 2007. *Sociologie du folklore brésilien et Études afro-brésilennes.* Paris: L'Harmattan. [1959]

Bateson, G. 1972. *Steps to an ecology of mind.* New York: Ballantine Books.

——— 1979. *Mind e nature: A necessary unit.* New York: EP Dutton.

Bender, C., and Mcroberts, O. 2012. "Mapping a field: Why and how to study spirituality". *Working Group on Spirituality, Political Engagement, and Public Life*, 1–27.

Brito, Gonçalves L. 2019. *O véu do Congá: Sobre três aspectos do conhecimento umbandista.* Rio de Janeiro: Gramma.

Camurça, Marcelo. 2003. "Espaços de hibridização, dessubstancialização da identidade religiosa e ideias fora do lugar". *Ciências Sociais e Religião* 5(5): 37–65.

Carneiro Da Cunha, *Cultura com aspas e outros ensaios.* São Paulo: Cosac Naify.

Carozzi, M. J. 1999a. "La autonomía como religión: La nueva era". *Alteridades* (9)18: 19–38.

——— 1999b. "Nova era: A autonomia como religião". In *A nova era no mercosul*, ed. M. J. Carozzi. Petrópolis: Vozes. pp. 149–190.

Clifford, James, and Marcus, George E. (eds) 1986. *Writing culture: The poetics and politics of ethnography.* Berkeley: University of California Press.

Csordas, T. J. 2009. *Transnational transcendence: Essays on religion and globalization.* Berkeley: University of California Press.

D'Andrea, A. A. F. 2000. *O self perfeito e a nova era: Individualismo e reflexividade em religiões pós-tradicionais.* São Paulo: Edições Loyola.

Danowski, D., and Viveiros De Castro, E. 2014. *Há mundos por vir? Ensaio sobre os medos e os fins*. Desterro [Florianópolis]: Cultura e Barbárie; Instituto Socioambiental.
De La Torre, R. 2011. "Les rendez-vous manqués de l'anthropologie et du chamanisme". *Archives de Sciences Sociales des Religions* 153: 145–158.
——— 2016. "Indo- and Afro-American religiosities, and circuits of new age". In *new age in Latin America. Popular variations and ethnic appropriations*, eds. R. De La Torre, C. G. Zúniga and N. B. J. Huet. Boston, MA: Brill, pp. 5–28.
Descartes, R. 1985. "Treatise on man". In *The philosophical writings of Descartes*, trans. J. Cottingham, R. Stoothoff and D. Murdoch. Cambridge: Cambridge University Press, pp. 99–108.
Featherstone, M. 1990. *Global culture: Nationalism, globalization and modernity*. London: Sage.
Ferguson, M. 1980. *The Aquarian conspiracy: Personal and social transformation in the 1980s*. Los Angeles, CA: Tarcher.
Gonçalves Brito, L. 2020. "Spirituality and ecology within the phenomenological realm of secularity". *Revista del CESLA* 26: 307–326.
Guerriero, S., Stern, F. L., and Bessa, M. 2016. "A difusão do ethos Nova Era e o declínio de seus estudos acadêmicos no Brasil". *Rever* 16: 9–39.
Hanegraaff, W. J. 1996. *New age religion and western culture: Esotericism in the mirror of secular thought*. Leiden; New York e Koln: Brill.
Heelas, P. 2006. "Challenging secularization theory: The growth of 'new age', spiritualities of life". *The Hedgehog Review* 8(1–2): 46–59.
Hervieu-Léger, D., and Hervieu, B. 1983. *Des communautés pour les temps difficiles*. Paris: Cerf.
Inda, J. X., and Rosaldo, R. 2008. *The anthropology of globalization: A reader*. Malden: Blackwell.
Ingold, T. 2002. *The perception of the environment*. Londres e Nova Iorque: Routledge.
——— 2005. "Human worlds are culturally constructed". In *Key debates in anthropology*, ed. T. Ingold. London: Routledge, pp. 112–118.
Latour, B. 1999. *Pandora's hope: Essays on the reality of science studies*. Cambridge, MA: Harvard University Press.
Luiz, R. R. 2013. "A religiosidade dos sem religião". *Ciências Sociais e Religião* 15(19): 73–88.
Magnani, J. G. C. 1999a. *Mystica urbe: Um estudo antropológico sobre o circuito neo-esotérico na metrópole*. São Paulo: Studio Nobel.
——— 1999b. "O circuito neo-esotérico na cidade de São Paulo". In *A nova era no Mercosul*, ed. M. J. Carozzi. Petrópolis: Vozes. pp. 27–46.
——— 2000. *O Brasil da Nova Era*. São Paulo: Jorge Zahar.
Meyer, B. 2014. "Mediation and the genesis of presence: Toward a material approach to religion". *Religion and Society: Advances in Research* 5: 205–230.
Odum, E. P. 2001. *Fundamentos de ecologia*. Lisboa: Fundação Calouste Gulbenkian.
Schutz, A., and Luckmann, T. 1973. *The structures of the life-world*. Evanston, IL: Northwestern University Press.
Steil, C. A. 2001. "Pluralismo, modernidade e tradição: Transfomações do campo religioso". *Ciencias Sociales y Religión/ Ciências Sociais e Religião* 3(3): 115–129.
——— and Carvalho, I. 2014. "Epistemologias ecológicas: Delimitando um conceito". *Mana* 20: 163–183.

Swyngedouw, E. 2011. "Whose environment? The end of nature, climate change and the process of post-politizacion". *Ambiente e Sociedade* 14(2): 69–87.

Toniol, R. 2015a. "Leaving the new age". In *Encyclopedia of Latin American religions*, ed. H. Gooren. Cham: Springer, pp. 1–2.

——— 2015b. *Do espírito na saúde: Oferta e uso de terapias alternativas/complementares nos serviços de saúde pública no Brasil.* Porto Alegre: Universidade Federal do Rio Grande do Sul.

——— 2017. "O que faz a espiritualidade?". *Religião & Sociedade* 37(2): 144–175.

Veer, P. 2009. "Spirituality in modern society". *Social Research* 76(4): 1097–1120.

Viotti, N. 2018. "La espiritualidad en América Latina". In *Religiones en cuestión: Campos, fronteras y perspectivas*, eds. J. C. Esquivel and V. G. Béliveau. Buenos Aires: Fundación CICCUS, pp. 233–236.

Viveiros de Castro, E. 2013. "The relative native". *Hau: Journal of Ethnographic Theory* (3)3: 473–502.

1 Exploring a fractal labyrinth

Figure 1.1 A river is never the same

Ethnographic travel consists of walking. A careful process in which one exercises attentivity (Ingold 2015). Nevertheless, it is neither a simple process with all the subsequent and distinguishable stages nor a temporally linear movement. Also, it is not a research technique by means of which one could arrive at a predetermined point, such as a map used to previously know the representation of the destiny.

Ethnographic travel includes being transported by different vehicles, which take the researcher from place to place. However, it is done by the very

DOI: 10.4324/9781003378853-2

legs, with the body, most of the time. Notwithstanding the necessity of preparation for the possible inclement weather on the way, and the necessary pack where one takes the materials for the path, there is always something acquired along the way. As well as something left.

The pack of concepts taken by the researcher for the anthropological exercise necessarily changes along the road. Interlocutors are also walkers in their life-worlds, who help the researcher in the task of transforming the concepts and finding better forms of sharing them, as well as the best places to find others. Then, the concepts brought in the anthropological pack are left behind, as seeds that now can be relevant to other people.

In a certain way, it is impossible to forecast when the concepts will change, when the route will change, because the path is only known when we pass through it. Serendipity is intrinsic to anthropological work amidst the ethnographic travel. Serendipity is the conceptual counterpart for the very life movement where the researcher drowns during anthropological research (Tsing 2015). Moreover, the acceptance of serendipity in the research corresponds to the understanding of the ways in which historical contingency and discovery influence the process of ethnographic travel. Serendipity characterises a research practice that depends on subjectivity and the complexities of human affairs, but it does not reflect a lack of scientific rigour (Rivoal and Salazar 2013; Hazan and Hertzog 2016; Martinez 2018). If ethnography is not seen anymore as a mere method of collecting empirical data during the fieldwork, it becomes more like an observational engagement with the world, including "long-term and open-ended commitment, generous attentiveness, relational depth, and sensitivity to context" (Ingold 2014: 384).

Amid the ethnographic travel, the observation requires the fluidity of the perception that follows the water dance on the mirror and bed of a river. Sometimes, the stillness of the surface hides the turmoil below, and numerous tiny beings that interfere with the flows, which are imperceptible without due attention. Social life often moves subtly. Ethnographic travel is a movement of openness through which the researcher finds the surprise, the exasperation and the prudence before the observed phenomena of the social world, like when we walk on a trail. Ethnographic travel is precisely the trail one follows in order to know the movement of life. And such a movement sometimes escapes from predefined categories, like when we look between the leaves on the soil of a rainforest and, taking off the leaves, the earthy sweet scent emanates whilst all sorts of insects reveal themselves underneath, working in unimagined symphony.

Evidently, ethnography is not considered here as a mere record of the lives of others. The description was only possible because there was a person who, walking in a labyrinth of concepts as a researcher, engaged deeply in the social worlds amid the intersubjective time where ideas, learning, dialogue and affection were shared.

Ethnographic travel is a formative and transformative experience. It is a process of walking that could never be reduced to a definite number of pages in the ethnographic text. Even if the final text captures the nuances of the travel, it is not the most important product because the process needs to be understood as a whole. The way can be described, but it could never be completely encapsulated only by the words. A dissertation during the journey of doctoral studies, for example, is rather a part of the broader process. The dissertation, as a prism of words, reflects and refracts the experience lived throughout the journey. The written text is thus a record of the transformative experience, as a part of the walking, and not as the final destination where to dock after the trip. Thus, the text is not the completion of the previously projected ethnographic travel, as well as the movement is not simply the effect of the mind that projects the action or the previous intentionality of the mind that plans the map by which it will follow the flow (see Ingold 2015: 133). It does not mean that the planning of the research or the report regarding the pathway travelled are not important, but rather that the complexity and profoundness of the travel cannot be deduced by the drawn map or the quantity of scenes of the narrative.

There are, evidently, various forms of narrating the ethnographic travel. However, all of them, implicitly or not, rewrite the modalities of the great narrative traditions. According to Benjamin (1968), two figures of storytellers have been relevant since the Classic times: the resident tiller of the soil and the trading seaman. While the first type cultivated the experiences of the place, finding the wisdom rooted in daily life, the second type brought the distant experiences of the people known through the travels. The two types of narrative construction interpenetrate because, frequently, the person who settled in a region once was the itinerant learner. The ethnographic travel also interweaves the two strands of experience in the narrative. Anthropology can also be made at home, although its epistemological assumptions will be different in this kind of research. But in ethnographic travel, the narrative arises from the experience of the person who travels alone, searching and bringing seeds from place to place. Ethnographic travel, so understood, is not only a method, but rather an effort of going to regions where nobody would go but us anthropologists. We are the people who navigate distances just to listen, dialogue and learn with people who live in these out-of-the-way places. And this is one of the great potentialities of Anthropology in these increasingly turbulent times of the Anthropocene (Tsing 2012; Gonçalves Brito 2021).

Coming in the labyrinth

My first ethnographic travel, during which I realised an exploratory fieldwork, took place in the second half of January 2018. Reading anthropological studies regarding spirituality, especially the writings of Amaral (1999, 2000), I realised that I could research the entanglements of spirituality and ecology in alternative communities. On the web, I had found lists of ecovillages in which I would spend some time as a volunteer. Volunteering is a common practice in ecovillages around the world. It is often realised through the contribution of a small amount for food and stay costs. Volunteering was a modality of immersion that enabled me to experience daily life in an ecovillage. Above all, I would have time to write notes, ask questions, interview people and so forth. I discovered and chose the ecovillage Arca Verde, situated in the region called Campos de Cima da Serra, 82 miles from Porto Alegre, main city of the state of Rio Grande do Sul, where I lived, by the time. I knew basically nothing about Arca Verde, nor the work of Comunello (2017), which was mentioned during my time at the ecovillage. I understood later that, in this case, there was a benefit to ignoring this dissertation because I did my fieldwork with my mind entirely curious about everything that was revealed along the way, including the enormous bushes of hydrangeas surrounding all the extension of the road in the region of Serra Gaúcha.

I arrived at the city of São Francisco de Paula approximately at half past nine p.m. on a Monday. As soon as I jumped out of the bus, Rodrigo and Amanda arrived, and took me in their small red car, through the way to Arca Verde Institute. There were plenty of araucaria and pines. The couple were psychologists, who lived in Minas Gerais, and were currently integrating into the community, in order to live there indefinitely.

Next morning, on January 16, I woke up at eight, and I talked briefly to Juliana (31 years old, pedagogue), who was taking her luggage to her car. She explained she was leaving, after the course "The nature of child", which took place during the weekend. I went to the kitchen for breakfast. Angelica introduced herself, and called me by name. She was a white woman, 30 years old, born in the countryside of the state of Rio Grande do Sul. I asked for an herb for the tea. She led me to the organic garden outside the kitchen. There were fruits and vegetables. We collected some leaves of lemon balm and I made the tea. Inside the kitchen, there were Felipe, Gil, Sinval, Amanda and Rodrigo. After breakfast, they began the weekly operational meeting, where they divided the tasks. Angelica asked me to introduce myself and I explained the subject of my research, in a general way, expressing my joy at being there. My contribution during the week would be: (1) to help Gil in the reform of the greenhouse and to clean the water tank of the lodging; and (2) to help Felipe in the construction of a vine structure for the bindweed. I washed the mug I used to drink the tea, following the shared norm of wash-what-you-use. The kitchen was the space where everybody met daily, at least for breakfast and lunch.

Gil and I went through the trail to the lodging, to complete our tasks. Gil was 72 years old. He had long white hair covering his shoulders, plastic sandals, and wore a green T-shirt and old blue jeans. He walked slowly and he spoke calmly. He graduated in communication from Catholic University, Porto Alegre. He travelled all around Brazil during his life, riding his bike. I helped him take the rotten wood pieces and rusty spikes off of the greenhouse structure. We nailed new spikes, covering the torn tarpaulin sheet with a new one. Gil was gentle and I noticed he enjoyed solitude. He did not like too much conversation, but he was patient when I asked simple questions about the tasks that, to his eyes, were simply daily procedures. We almost completed the task of the day and decided to finish the following day. I then cleaned the room where I left my backpack and walked back to the kitchen.

During lunch, I saw other people I had not seen yet. I was sitting on the porch, looking at the green of the trees and the sunny limpid sky, in silence, while I ate. Then, I walked with Juliana, as a farewell. She photographed the trash shed, where people separated the different residues and kept the material for the bioconstruction.[1] Inside the shed, there was a can full of plastic and plastic bottles. The plastic was inserted in the bottles, which became a kind of very tough brick, used for the construction of the structures of walls.

Juliana and I passed by the side of the lake, where naturist bathing is allowed. We walked through the trail that led to a crossing. We found two signs with two words: "labyrinth" and "temazcal".[2] We proceeded to the labyrinth, which looked like a circular and spiral path of stones. Under the araucarias and pines, in the middle of the labyrinth, we stayed for a few minutes. This was the threshold of the labyrinth of an environmental anthropology whose main research fields amid the ethnographic travels were always the ecovillages.

Ecovillages are places where the ecological apprenticeship happens through the senses, lived experience, affects, and also spirituality. The research of Comunello (2017) approaches the relevance of materiality and engagement in the processes of apprenticeship that collide with the onto-epistemic great divisors of Euro-American modernity. Moreover, Comunello also demonstrates how materials are entangled during processes of formation of the ecological subjects in the context of ecovillages.

My study shares with Comunello (2017) the theoretical and methodological commitment regarding the exploration of the paths of ecological epistemologies, which disrupt the representational perspective. According to ecological epistemologies, knowledge is produced through the experiential engagement in the world. They also return to materiality, breaking down the dichotomies of Cartesian science. Finally, and no less importantly, they postulate that all beings share one environment, repositioning the human as symmetrical to the non-human (Steil and Carvalho 2014).

My approach is different because my research is immersed in different trails, even though Comunello and I have been partially navigating the same river, which, in fact, has never been the same. My study emphasises different conceptions on spirituality that could be understood as the experiential counterpart for the ecological epistemologies. The ethnographic text is built upon a critical description, highlighting how such spiritualities and ecological epistemologies are entangled in the lived experience of dwellers of four ecovillages.

On the other hand, my dissertation intends to delineate these reflections from some possible threads of what could be an environmental anthropology. Ecological and environmental anthropology has been constituted as a subfield of the anthropological discipline in which one researches the relations between human and environment. Following the current unfoldings of cultural ecology, the approach of ecological and environmental anthropology encompasses the "totality of relations existing between persons and their environments and privileges neither genetics nor culture in explanations of human action and perception" (Ingold n.d.).[3] Environmental anthropology is inherently interdisciplinary, creating a dialogue with environmental humanities, cultural geography and science and technology studies. Some questions, such as the cultural apprenticeship through which humans share their understandings on the world they inhabit, are very close to the issues explored by environmental education, another transdisciplinary field.

I am more interested in emphasising the term environmental anthropology in the research vocabulary of Brazilian anthropology than delimiting the specificities of the space an environmental anthropology may occupy, in terms of academic specialty.[4] I would rather use the term as a point of convergence between an anthropology of life, an anthropology of experience, an anthropology of space and place, a phenomenological anthropology, a multispecies anthropology, an anthropology of the Anthropocene, and other correlate anthropologies. Thus, I do not intend to delimit another field of investigation that fragments an already fragmented discipline, but rather integrate various problems explored by specific anthropologies that seem to increase their potentialities when they come together.

My premise is the understanding that all anthropologies share a common epistemological ground – difference, differentiation, otherness, alterity and diversity – instead of being defined by their specific topics of research. Thus, this study does not shy away from the topics explored by different anthropologies when such topics arise from the very research. The environmental anthropology delineated above is defined as the methodological space in which a little part of the complex human engagements with their environments is encompassed.

Exploring an ecovillage

On the morning of the third day, Angelica and I went for a walk, so she would show me the physical spaces of Arca Verde. Angelica was wearing a very long almost-touching-the-ground skirt. Her long blonde hair was tied by a cotton headband on her forehead, under which attentive eyes were visible. First, she took me to the yurt[5] where the dwellers usually held cultural moments and other practices of dialogue and sharing. Exceptional meetings and feedback reunions among volunteers and dwellers happened there. Outside the yurt, there was a space where people should leave their shoes before coming in. There was also a shelf with magazines and books. On the cover of a magazine, I read "Ecology is spirituality". Later, when I returned there to pick the magazine up and read it, it was not there anymore. Serendipitously, my eyes had glimpsed the headline and, in a few seconds, one premise of my thesis was revealed. Later, I heard and read the same notion in other places, until I understood that some modalities of spirituality and ecological thought are like two rivers that drain into the same ocean.

Inside the yurt, there was an infographic in the shape of a flower, whose bud integrated the ethics of Arca. It was the permaculture flower, whose petals linked (1) the built environment; (2) tools and technologies; (3) culture and education; (4) health and spiritual well-being; (5) finance and economics; (6) land tenure and community governance; and (7) land and nature stewardship. Each item involved specific interrelated activities and processes that integrated the permaculture principles. I discovered that 13 children were already born in the ecovillage through humanised birth. I also asked Angelica about the association between Arca Verde, Casa Brasil and GEN (Global Ecovillage Network) and she told me that, at that moment, there was no project articulating the networks closely.

Angelica told me that Arca Verde aimed for food sovereignty, and that it was still a real goal. Angelica also demystified the idealistic image of ecovillages and other ecological communities – eating the cultivated food is not that easy. An example was the cultivation of carrots. For some time, the carrots did not grow sufficiently large. Only after many trials did they present a reasonable size. Caring for the earth was constant. It was not only opening the soil to leave seeds. It was necessary to pay constant attention to the growth of the different species, so they would help each other. The biodiverse cultivation required accurate knowledge on the life cycles of each plant.

Angelica and I also talked about the principles of coexistence in the ecovillage, based on non-violent communication. People were advised to avoid gossiping and expression of negative opinions regarding someone to another person. Angelica called this non-triangulation. When an uncomfortable situation happened, the involved people were advised to dialogue.

After the guided tour, I ran to help Gil with the greenhouse again and we finished its ceiling. Then, I withdrew all the humid and wet things. I also removed small leaves of grass with the rake and little spiders and earthworms scattered, frightened beneath the layer of sawdust that covered the ground.

Afterwards, I helped Sinval who was storing the tools he used to repair the pedal boat, which was floating on the lake. He had used cords tied on the trees as pulleys. He, who was more than 60 years old, managed to remove the pedal boat from the lake without physical force, using a trick of mechanics.

Everything surprised me.

At night, I nurtured myself with food and went to the co-living space (called redondo). Above the threshold, there was a sign with the following Hopi idea: "We are the ones we've been waiting for". Franco arrived at the space and greeted me, mentioning my situation as a researcher. He had a 30-year-old face, and had worked with communication and arts as a young adult. He had been in the ecovillage four years before and decided to become a dweller. He became one of the mentors of the course "Dragon Dreaming", whose objective was the elaboration of creative collaborative projects. We started a dialogue because I noticed he was spontaneously gentle, as if he had worked inside himself for self-transformation. During those days, I understood the gentleness, kindness and receptivity expressed by those people could only be associated with spirituality. He confirmed my impression when he told me he has been practising yoga since he was 15.

Franco told me of his experiences of observing earth and perceiving moving life in each extension of soil. That reminded me of the ideas of Ingold. Franco had not read Ingold, but he said:

– In this community, I am a set of meetings, like these socks, like this table, like you. There is a community in me and there is a community in you. Many cycles are happening within all these communities.

The person, the things, each one, presented themselves like a gathering of threads of life (Ingold 2008). Each being and thing could be understood as a thread of lines, a community in which such threads meet. The meeting of two persons was then a gathering between communities. Franco understood our dialogue and social life in the ecovillage like this meeting of communities, including the dwellers, the visitors and volunteers. Franco also understood that the spiritual connection of the group was based on the "exercise of supporting life" because "favouring life in its different aspects was the strongest spirituality of the group". The ecovillage was like a kind of laboratory where people shared the collective apprenticeship of learning how to be attentive to life. Cultivating a sensitivity to nature would teach humans that humans and nature are entangled by the same patterns of the life that moves underneath and on the surface of earth. Franco said: "As above, so below". With the aphorism, he created an analogy of the movement of seasons, cycles of plants, the biodiversity of species and human movements, their cycles and diversity among people. Finally, describing social life in the community, he asserted: "Everything is a temporary node and a process of continuous flow".

The engagement with the world of agroforestry

Biodiversity of species is a basic premise of agroforestry, a practice that aims at the integration of different kinds of cultivation in the same space, usually called system. The agroforestry system of Arca Verde was affectionately known as Rosaf, a term that united the term "roça" (how Brazilian farmers usually call the cultivated land) and the suffix "saf", the agroforestry system. There were more than a dozen plats, whose extension was approximately 10 metres each. The disposition of plats followed the same structure. One plat was dedicated to woody perennials and the next plat was the space for seasonal cultivation of alternate species, especially fruit trees. The perennials were pioneer trees that initiated the system and also plants that provided shadow and nutrients for the seasonal species. The carmerão, also known as capim-elefante (*Pennisetum purpureum Schum*), whose fibrous stalk looked like the corn, provided nutrients to the growing vegetables and fruit trees. I spent the mornings cutting the stalks of carmerão and leaving them around the plats of potatoes. The stalks were left upon the ground, around the leaves sprouting out of the earth, fertilising the soil and protecting it from the sun.

I was taking care of the Rosaf, whilst Alexandre dumped sand from a wheelbarrow. I asked him to share some information regarding the agroforestry. Alexandre had just arrived from a trip with his partner, Liriane. Both were in their early twenties. Alexandre also took care of the agroforestry system, which was the principal focus of my volunteering period. He said that agroforestry refers to a method that aims to respect the biodiversity and interrelation between plants, in order to emulate the diversity of a forest. The hands of the agriculturalist speeded up some process that happens naturally in the forest – in the forest, for example, the fallen leaves of plants provide nutrients for other plants and, in the agroforestry system, the agriculturalist would speed up this process, cutting some plants at the appropriate moment and providing the compost for other plants. According to Alexandre, some trees were pioneers, "placental" because they helped to prepare the soil, providing nutrients, shadow for the growing plants and also firewood, after they finished their function in the system.

Rosaf was a forest in miniature. Uvaias, cúrcuma, ipês, arrudas and other trees coexisted in the same space. The cultivated biodiversity was oriented to the integration of plants. Some trees captured the nitrogen and distributed it underneath the soil through the roots. According to Alexandre, the external products that some conventional farmers use to fertilise the soil, such as nitrogen and phosphorus, are naturally present in the soil through the action of trees. Each plant, as living organisms, if their cycles are observed and respected, grows without any action of pesticides and fungicides.

Felipe (a 21-year-old man who was born in São Francisco de Paula and lived in the ecovillage for three years) mentored me during my stay as a volunteer and answered my questions. He explained that the Rosaf was not only an integrated system of cultivation, but also a system with a particular energetic flow. An agriculturalist would become part of the flow when they are taking care of the land. The system needed constant and rhythmic touch in order to thrive. Cultivating was not simply putting the seeds on the soil, but rather taking care of the earth, dancing with life.

On that day, during our dialogue after lunch, Marcos told me about his spirituality. His favourite practice was laying on the ground and staying still, observing the movement of the trees. Staying near the stream. Walking in the forest. These practices enabled him to feel the connection with nature. Marcos understood the search for the sense of such connection was inherently human, for humans have always been in contact with nature, during all the history of evolution. He mentioned the numberless papers that demonstrate the beneficial effects of the contact with nature over human health. And he, who walked barefoot even in the winter, was very aware of the importance of his meditative practice in connection with nature to the immunity of his body.[6]

Marcos complemented his thought: – But the human being is also a social being. So we do have this necessity of social contact. Maturana wrote that many illnesses are related to the interference on the biology of love. We have as mammals an emotional background which is amorousness, the acceptance of the other to coexist. When this basic element is broken, the biological harmony is broken and then the disease appears. According to Maturana, another basic element of the social relations of beings among them and their environment, during all those billion years of evolution, is cooperation. When we culturally interfere with ideas of competition and domination of others, we break the biology of love and get sick.[7]

Marcos was presenting to me a notion of spirituality as the search of connection with nature. Nevertheless, he also understood that such a perception of the "structural coupling", the reciprocal changes influenced by the contact with beings among them and the environment, can also be cultivated. Furthermore, Marcos understood that cooperation and amorousness were the natural ground in which humans, society and environment were entangled. Thus, the sense of connection with nature is a result of evolution, positively impacting human health and enabling the maintenance of life, which is intimately tied to the lived environment. According to the concept of Maturana and Mpodozis (2000), the history of the diversification and maintenance of life of organisms cannot be reduced to the fight for survival of the fittest, in the context of a competitive logic.[8] The spirituality of Marcos, grounded in the notion of the search of connection between human and environment, was existentially rooted in an ecological epistemology.

Marcos was a 39-year-old biologist. Years before, he had travelled to Santiago (Chile) looking after Maturana. Marcos thought he would find him at the Instituto Matríztico, founded by Maturana and Ximena Dávila, in 2000. Marcos really appreciated the ideas of this biologist/philosopher. Marcos did not become a professional biologist and, for one year, lived in Germany, where he visited many alternative communities of GEN. In 2010, when Marcos returned to Brazil, he became a dweller of Arca Verde. Through our dialogues, Marcos brought important elements that helped me in the process of rethinking and transforming the problems of my research. When I revealed the general theme of the research, he suggested the replacement of the notion of climate change for ecological crisis and that helped my comprehension of climate change and other anthropogenic changes over the ecosystems as an aspect of a much broader ethical, philosophical and existential problem.

We talked about spirituality and ecology, perception of the environment and consciousness, whilst walking through the forest of araucarias. Marcos was holding Lisbela, his newborn little daughter, on a sling (a kind of cloth that crossed the dorse and shoulders; used by parents to take their snuggly children). When she slept, we decided to work on the production of shiitake mushrooms. In a glade, there were wood logs forming a structure covered by transparent tarpaulin. There was also a desk and a rocket, a kind of portable stove made of aluminium cans. Our task consisted of melting the wax and covering the open holes in the logs, where fungi and sawdust were implanted. The organisms would feed from the wood and, in the propitious moment of reproduction, lift their reproductive organs, which are the mushrooms. We talked about the lack of ethical responsibility and political challenges in colonised countries that prevented a more coherent and effective answer to this apparently evident global problem, the ecological crisis, in all its aspects.

Marcos said: – In the colonised countries, people are more individualistic, they look for liberty and private property and do not care for inequalities. But the crisis is happening. The Amazon is being destroyed and only 5 percent of rainforest remains. Rarely a clear river in urban environments. Air pollution. Toxic waste. Animals facing extinction. In the cities, people only feel bothered when tap water is lacking. If there is energy, water, food, people do not care. Even though we have information, it is too much information and people are anaesthetised. Many people know what happens, but this is not sufficient for a shift in lifestyle. It is a cultural issue too.

I returned: – Despite the information, the common people do not grasp the functioning of a hydroelectric plant, how water arrives to their houses, or all political and economic processes involved in the construction of a hydroelectric plant, deforestation, mining and monoculture. The person goes to the supermarket and sees that enormous beautiful vegetable, with no spots. Maybe, that person does not care for the poison that became part of the substance of the lettuce. The building of a hydroelectric plant, the soy monoculture and deforestation are very distinct things, but all these phenomena are embedded with the consumerism logic, the logic of capitalist accumulation. These phenomena are part of the same system. It is not obvious to many minds.

Marcos added: – In the economic theory, all of these are externalities. Pollution, waste, the damage of the materials. The environment is only an externality that does not matter. Profit matters.

In certain trends of economic thought, particularly those that follow a notion of "weak sustainability" (Carvalho 2008), the environment is an externality out of the scope of the internal logic of companies' decision-making whose general aim is profit, even if the profit comes from the compulsory displacement of populations, pollution of rivers or deforestation in Indigenous territory. According to that logic of the capital and resources expropriation, the damage done to those who do not participate in the profit interest of the companies is also considered an externality. Nevertheless, even the internalisation of externalities, through financial indemnity to those who were injured by the entrepreneurship, cannot prevent the crassest disruption of ecosystems and the impact of the projects on the flows of human and non-human life.

A stoic anarchist

Aldo was 40 years old. He was born in Criciúma, Santa Catarina. He studied psychology. When he first heard of Arca Verde, he was living in Florianopólis. He worked as an animal caretaker in a farm where many young travellers passed in their quest for alternative modes of living. The year was 2007 and Arca Verde was still situated in the town of São José dos Ausentes, where the frosty weather practically prevented the cultivation of food. Aldo participated in the first group of volunteers, in 2008. In 2009, Arca Verde moved to São Francisco de Paula, where the warmer weather allowed for some cultivation. Four years later, Aldo returned to the ecovillage with the idea of staying there. His main motivation for living in community was his generalised unhappiness.

Aldo was raised by a catholic family. As a teenager, he joined gnostic groups. Reading Krishnamurti, he began to experience an individualistic spirituality. In his own words, Aldo freed himself from brainwashing. He told me, with a smile: "The founder of modern gnosis, Samael Aun Weor, studied lots of Esotericism, made a salad, invented a little bit and created the Gnosis. When I left the institution, I read Gurdjieff and I noticed that Samael copied two pages of one Gurdjieff's book, word by word, without any reference".

When I met Aldo, his spiritual practice was the state of attention. A state of attention he could exercise every time. An attention opened to everything that appears. "Each thought, each emotion, each movement of the external world", he said, "to distinguish what I am from what I am not". Such a practice was specially inspired by the philosophy of Krishnamurti. In this state of attention, the perceived world is not conditioned by a mind that judges, but rather by a mind that perceives entirely with the state of presence.

Along with the other teachings of Krishnamurti, Aldo also commented on the interdependence between individual and society. Such a concept challenged the infamous dichotomy. Aldo understood, through his readings of the Indian philosopher, that the aggressive and fearful thoughts and deeds of the individual had the same nature of the violence, fear and aggressivity of the collective. Thus, individual transformation would have the same nature as social transformation. He said to me:

– In the mass movements, politics and militarism, the need for changing society is the dominant idea. The individual needs to adapt himself. The idea of changing society without changing the individual. But when one looks toward changing the individual through propaganda, movies, news and internet, it is brainwashing. One tries to mold these individuals so they can support certain collective action. Even political power considers the importance of changing the individual in order to create an effect on society. The ones who hold the political power try to feed and strengthen the ideas of patriotism and nationalism, which will favour the centralised power and cement this false idea of democracy, as if the vote was a possible way to change things, as if the state was necessary. The state knows the necessity of the individual action. But the population does not have the vision of itself as a force of change in society.

While I shared my analysis, Aldo shook his head as a sign of agreement: – There are three points in this. First, the interdependence of the individual level and the social level is such an obvious reality that power holders use it, according to a logic of political power, to dominate the masses and maintain the privileges of the few in detriment of the necessities of the population. Second element, the political elite, the powerful agents of the state, who are interested in keeping the status quo, know they need to act on the individual level in order to make people think there is a state outside them. And that this entity and its invisible hand are inherently kind. The state (and the political-economical structure that sustains it) is frequently understood as "an entity with a life of its own, distinct from both governors and governed" (Asad 2004: 281). The state is understood as a disembodied and self-existent entity, floating upon and beyond people, society and economy. Although the state is rather the effect of the actions performed by bureaucrats in their daily practices of organisation of space and time, supervision and surveillance (Mitchell 2006). And these bureaucrats are, evidently, people. Third aspect, the very concept of society as a thing outside the individuals, larger than them, and bigger than the sum of the parts. All these disjunctions contribute to the common perception of one person's action in daily life as inexorably separate from the action of other people in their lives as individuals. And this separate mind, individualised like this, tends to reflect on social problems as external things. Such a mind does not feel responsible for the collective problems. On the other hand, people who understand interdependence can act to transform society, even if they see it as something exterior. Such a person can feel insufficiently strong to fight against this apparently much bigger thing. But if we bring the concept of Krishnamurti, according to which the individual transformation is social transformation, what can we say?

Then Aldo also shared his reflection: – The largest the number of people who achieve self-transformation, the broader the influence of a more elevated consciousness in society. To me, there is nothing mysterious about this. And I would like to emphasise that the way for achieving a higher consciousness is not the way for interfering in society. I really think that when the person is full of good intentions and willing to interfere in society, chances are they will not obtain good results, if the person has not achieved this needed state of interior quality. I see this in the action of some feminists, who become aggressive and attack a portion of society. They are very well-intentioned. They want equality. But they end up feeding rancour. Same way, the well-intentioned permaculturalist full of good intentions. If his action comes from an inner state of fear and aggressivity, it will never generate a good fruit. So, the idea is going within and solving all your problems and, then, detaching from the mind and letting the action rise or not. And simply be at peace. But maybe this peace will not be complete because the love people feel for others will lead them to suffer in terrible situations. The higher the inner quality of people who are not willing to change society, the better their influence in society. The transformation of the individual is more important than the transformation of society because the problem of society is the individual. If the individual does not heal, keeping their violence, fearfulness, greed always avid and insatiable, the willingness to change society is insufficient, if this movement of changing society comes from such an individual, with all those problems. Why would the individual who healed be good to society, even if they do not act? They will be good for their non-action. Taoism highlights this idea of non-action. In non-action, the person is not being violent, fearful, but only emanating love. Thus, this still, peaceful, non-violent and fearless person is contributing more than the violent and fearful person who are willing to change society. Even though the contribution of the still person is negative. Such a person is not feeding the fire. But that does not mean they will remain negative, for the action may surge. Other way would be going within with the intention of helping others. Thus, the person can act by love for others and influence positively. But only if this person goes within and, dissatisfied by what is seen inside, solves the problems attentively, before willing to interfere in other people's lives.

Our dialogue encompassed the notion of spirituality as an inner transformation that resonates in the lived world, generating social transformation. Realistically, the person who experiences such a transformation can only change the social world in which they live, and not the world in general. Such a notion becomes clearer when the interrelation between individual and society is understood beyond the illusory dichotomy. Nevertheless, according to Aldo's ideas, the transformation can happen in the milestone of free will, autonomy and horizontality of social life. Such ideas are the flowers of an anarchist political philosophy. Aldo told me what he understood as anarchism:

– To me, anarchism is not a mess. Anarchism is the absence of hierarchy. Monarchy, capitalism, army, communism and religion have hierarchy. The absence of hierarchy does not mean absence of rules, but the rules are created in a horizontal environment. Not by the few who create rules for the mass to follow. Here in Arca, for example, there is no person who owns and rules.

I act in the sense of defending this. Even though there is a force here that aims to the contrary, Arca is anarchist.

Throughout this text, the speech of interlocutors is marked by spaces above and below, typed in the same font, with a few common exceptions of literary writing, in which multiple voices are introduced by quotation marks. In doing this, I intend to demonstrate that people's analyses are not mere data used to corroborate a thesis.

Evidently, the concepts and experiences of others also nurture and corroborate the thesis, but they have a proper epistemological foundation. Such a recognition requires the exercise of an epistemological humbleness, which enables the refusal of a hierarchy between the researcher and the so-called informant. It does not mean that ideas of interlocutors are always irrefutable, correct and precise. In many contexts, they are exactly the contrary. Epistemological humbleness stems from the recognition of interlocutors as "full-fledged personalities" whose experiences and ideas can be contextually located if we "acknowledge their history, and position them in a particular social setting" (Bruner 1993: 327). The methodological choice for keeping the names of interlocutors is also derived from such a recognition. Thus, their concepts and experiences, albeit different from anthropological theory and analysis, are considered on the plane of a symmetry between modalities of knowledge. Their validity is recognised epistemologically. Finally, people do not live passively without reflecting on their experiences. On the contrary, they live, reflect on their experiences, and act according to the conceptions they build from the lived experience (Turner 1986). I call this lived knowledge. And I am entirely convinced that listening to people creates the possibility of describing their lived knowledge and knowing the social worlds in which they inhabit.

The researcher looks for sharing intersubjective time with the interlocutors through an engagement in the dialogue, refusing ascendancy of the scientific knowledge cultivated by the researcher. The symmetry between anthropological theory and the concepts introduced by interlocutors can only be reached by the recognition of coevalness of the fieldwork experience (Fabian 1983). The coevalness is expressed by the ethnographic text where the ethnographic present is avoided. The ethnographic present is the language used to build the aporetic temporal distancing between the moment of the field and the moment of the analysis. The paradox is the fact that such a distancing is expected to create certain neutrality through the usage of present-tense verbs, the division between participant observation and anthropological analysis and the obnubilation of the past time when the research was done.

Nevertheless, the ethnographic effect, according to Strathern (1999), involves observation and analysis as two interwoven – not separate – moments. These two moments are not two subsequent stages in a supposedly linear time that could be identified in the research process. In my ethnographic research – which I would like to name ethnographic travel – we observe and analyse simultaneously. Observation and analysis, the dialectics of immersion and movement, consist of "a relation which joins, the understood (what is analysed at the moment of observation) to the need to understand (what is observed at the moment of analysis)" (Strathern 1999: 6). Therefore, every ethnographic narrative that introduces the observed field can contain anthropological theory, and vice versa, in a form of mutuality that does not exclude the different aspects of research.

In this dissertation, inspired by the critical description in multispecies ethnography, the analytical reflections are presented alongside the ethnographic narrative of my fieldwork, emphasising the concepts of my interlocutors, describing the details in the world and raising bigger questions from the details, as I have heard from Tsing (2013, 2015). The textual relation between theory and ethnography is, consequently, fractal. Theory always contains the concepts and experiences lived through ethnography just as the description contains theory.

The premise of fractality as a core tenet of this dissertation appears recursively alongside the chapters. Thereby, the experiences and conceptions in the social lived worlds are combined as parts that contain the whole. As if the first verse of a poem expressed the leitmotif of the whole set of strophes in the poem. As if the preliminary seconds of a song unveiled the prelude of the entire melody. Thus, the totality of the reflection emerges from the details of the experiences lived in the different ecovillages, alongside the ways and movements through which those people passed. However, such a whole that emerges from the parts and the recursivity of the whole through the parts cannot be grasped if we look from above. I am not providing a map here. I am advising about a reading path in which the forms of repetition of experiences and conceptions will only be noticed alongside the path. Hence, I invite you to come into this fractal labyrinth, where the part is whole and the whole is part. There is nothing more coherent with the premise of many people with whom I dialogued: the notion that society is in the individual whilst the transformation of the individual is equivalent to social transformation. Or the notion that humans are in nature as nature is in the human.

Notes

1 Bioconstruction can be understood as a set of building techniques that used the materials existent in the region – such as sand, animal excrement, straw. There are many forms of building walls and houses structures: *Minke* bricks, superadobe, adobe and so forth. In the ecovillage, the fabrication of bricks was a daily activity. The materials used were red earth, sand and water. The mud was inserted in wooden moulds. After exposure to the sun, the cob became square, and the result was firm bricks. To focus on these technologies and "flows of materials" (Ingold 2010) would provide enough content to a complex and study on the anthropology of technique.

2 One of the signs indicated the place where people held rituals of temazcal. A neoshamanic rite, adapted from Nahuatl tradition. Temazcal is the sweat tent,

"the experience of being close to the 'uterus of the mother earth' – a ritual of purification and rebirth" (Comunello and Carvalho 2015: 93).

3 This entry was written by Professor Ingold and can be found at https://www.discoveranthropology.org.uk/about-anthropology/specialist-areas/ecological-environmental-anthropology.html. Available on August 4, 2021.

4 As far as I know, Paul Little sometimes used the term environmental anthropology during lectures and put it on his Lattes curriculum, even though he wrote specifically about the correlated field of political ecology (see Little 2006).

5 Yurt is a circular building made of bamboo and a tarpaulin sheet, which can be undone and redone – it was a habitation used by nomadic peoples in Central Asia for at least 3,000 years.

6 According to Carvalho and Steil,

> The experience of being in natural places and carrying out ecological activities and rituals linked to nature, results in educational processes which develop skills and re-enforce expectations of authenticity, transcendence, health, well-being, healing of the self and the planet, aesthetic pleasure and moral correctness.
>
> (2013: 115)

7 The reference mentioned by Marcos can be found in the following book: Maturana, H. R., and Verden-Zöller, G. 2003. *Amor y juego: fundamentos olvidados de lo humano, desde el patriarcado a la democracia.* JC Sáez editor.

8 Maturana and Mpodozis (2000) challenged the common concept of natural selection, proposing the concept of natural drift. The authors argued that "the diversification of lineages in living systems does not course in a competitive dynamic through the survival of the fittest, but it follows the course of the survival of the fit in the conservation of autopoiesis (living) and adaptation" (289).

References

Amaral, L. 1999. "Sincretismo em movimento – O estilo Nova Era de lidar com o sagrado". In *A Nova Era no Mercosul*, ed. M. J. Carozzi. Rio de Janeiro: Vozes, pp. 47–79.

——— 2000. *Carnaval da alma: Comunidade, essência e sincretismo na Nova Era.* Rio de Janeiro: Vozes.

Asad, T. 2004. "Where are the margins of the state?" In *Anthropology in the margins of the state*, eds. V. Das and D. Poole. Santa Fe: School of American Research Press, pp. 279–288.

Benjamin, W. 1968. "The storyteller". In *Illuminations: Walter Benjamin essays and reflections*, ed. H. Arendt. New York: Schocken.

Bruner, E. 1993. "Epilogue: Creative persona and the problem of authenticity." In *Creativity/anthropology*, eds. S. Lavie, K. Narayan and R. Rosaldo. Ithaca, NY: Cornell University Press, pp. 321–334.

Carvalho, I. C. M. 2008. "Educação para sociedades sustentáveis e ambientalmente justas". *REMEA-Revista Eletrônica do Mestrado em Educação Ambiental*, pp. 46–55.

——— and Steil, C. A. 2013. "Natureza e imaginação: O deus da ecologia no horizonte moral do ambientalismo". *Ambiente & Sociedade* 16: 103–118.

Comunello, L. 2017. *Aprendizagem e espiritualidade em ecovilas: Quando "o universo todo ensina".* Porto Alegre: Pontifícia Universidade Católica do Rio Grande do Sul.

Comunello, L., and Carvalho, I. 2015. "Ecovilas: Aprendizagens, espiritualidade e ecologia". *Avá Revista de Antropología* 27: 81–99.

Fabian, J. 1983. *The time and the other: How anthropology makes its object*. New York: Columbia University Press.

Gonçalves Brito, L. 2021. "Futuros possíveis dos mundos sociais mais que humanos: Entrevista com Anna Tsing". *Horizontes Antropológicos* 27: 405–417.

Hazan, H., and Hertzog, Esther. 2016. *Serendipity in anthropological research: The nomadic turn*. London e New York: Routledge.

Ingold, T. 2008. "Bindings against boundaries: Entanglements of life in an open world". *Environment and Planning A* 40(8): 1796–1810.

——— 2010. "Bringing things to life: Creative entanglements in a world of materials". *Realities Working Papers* 15: 1–14.

——— 2015. *The life of lines*. New York: Routledge.

Little, P. 2006. "Ecologia política como etnografia: Um guia teórico". *Horizontes Antropológicos* 12(25): 85–103.

Martinez, F. 2018. "The serendipity of anthropological practice." *Anthropological Journal of European Cultures* 27(1): 1–6.

Maturana, H., and Mpodozis, J. 2000. "The origin of species by means of natural drift – El origen de las especies por medio de la deriva natural". *Revista Chilena de História Natural* 73(2): 261–310.

Mitchell, T. 2006. "Society, economy and state effect". In *The anthropology of the state: A reader*, eds. A. Sharma and A. Gupta. Oxford: Blackwell Publishing, pp. 321–334.

Rivoal, I., and Salazar, N. B. 2013. "Contemporary ethnographic practice and the value of serendipity." *Social Anthropology* 21(2): 178–185.

Steil, C., and Carvalho, I. 2014. "Epistemologias ecológicas: Delimitando um conceito". *Mana* 20: 163–183.

Strathern, M. 1999. *Property, substance, and effect: Anthropological essays on persons and things*. London: Athlone Press.

Tsing, A. 2012. "Unruly edges: Mushrooms as companion species: For Donna Haraway". *Environmental Humanities* 1(1): 141–154.

——— 2013. "More-than-human sociality; A call for critical description". In *Anthropology and nature*, ed. K. Hastrup. New York: Routledge, pp. 27–42.

——— 2015. *The mushroom at the end of the world: On the possibility of life in capitalist ruins*. Princeton e Oxford: Princeton University Press.

Turner, V. 1986. "Dewey, dilthey, and drama: An essay in anthropological experience". In *The anthropology of experience*, eds. V. Turner and E. Bruner. Urbana/Chicago: University of Illinois Press, pp. 33–44.

Interregno

On ontological insecurity in Porto Alegre caused by the fossil fuels crisis, which resembled an Armageddon

Figure Interregno 1.1 A storm

The standstill of truckers, as a protest to the exorbitant taxes imposed on fossil fuels like gas and diesel, in the context of the financial crisis of the largest Brazilian petrochemical company Petrobras, caused, for many days, a collective panic. People were frightened by the idea of food scarcity, the roads were blocked and the trucks held much of the food supply. The country was immersed in this scenario of uncertainties, in part due to the impeachment/coup d´etat that deposed President Dilma Rousseff two years before. People were fearful, worried about the future of Brazilian democratic institutions.

DOI: 10.4324/9781003378853-3

The corruption scandals at every level of legislature, revealed through social media, contributed to the poignant polarisation of the political debate nationally.

News circulated in the city of Porto Alegre, with the rumour that food supply was lacking. The prefectures of many cities, Porto Alegre included, declared a state of emergency. Public transportation was reduced and activities and events at public and private universities were cancelled. On Thursday, May 24, 2018, the workers of an elite supermarket in Porto Alegre witnessed the crowd who shoved and thrust before empty shelves.

On Saturday 26th, I took my bike and saw endless lines in front of gas stations. When I arrived at the organic farmers market, Feira dos Agricultores Ecologistas, there were not even any bananas. It was 11 a.m. There was an apocalyptic aura in the farmers market too. Scared people took all the fruits and vegetables. An agriculturalist sold all the vegetables and returned twice to his farm in a close city, Eldorado do Sul, in order to bring more. I also met Mr. Juarez, a guardian of biodynamic rice seeds, who had only three small sacks. He rarely came home with so few. I also spoke to Márcia and Marinês, who sold whole bread. We discussed how apocalyptic ideas frighten people. Márcia showed me her texts with her husband. The husband asked how everything was there and her answer was: "It looks like the Armageddon, the end of the world". On the previous day, worried by the news, she had even thought of producing double the number of loaves to account for food shortages.

The report above is the leitmotif of my reflection on some texts I read during the course Politics of Nature, offered by Isabel Carvalho and Carlos Steil (Federal University of Rio Grande do Sul-Brazil). The notion of securitisation (Warner and Boas 2017) enables the understanding of such a curious phenomenon of entire populations who are so fearful they are capable of stocking food before the uncertainty generated by the possibility of scarcity of food in the markets' shelves.

In resonance with the so-called Copenhagen School, Warner and Boas (2017) suggest that the notion of securitisation implies the framing of a public issue as an existential threat through the production of calamitous discourses. Such discourses use the rhetoric of disaster, emergency and crisis, figuring a kind of threat that challenges certain social assemblage. This is turn justifies urgent actions of security and opening a "policy window" easily used to insert specific agendas that would not have any force without the scary situation framed by the very discourses.

In the case of the official discourses concerning climate change, examples include the security framing of Prime Minister Tony Blair and Foreign Secretary Margaret Becket during the Copenhagen UNFCCC Conference of December 2009 or the Dutch Delta Commission framing of flood security in 2008. This policy rhetoric of crisis amplifies the sense of catastrophe and aims "to legitimise coercive measures that are impossible in normal times" (Warner and Boas 2017: 206). The alarmist narratives on climate change mobilise people to take part in the catastrophe, but also fuel "public disengagement with climate change and promote a sense of fatalism or scepticism" (Warner and Boas 2017: 215).

There are some relations between the catastrophic discourses on climate change and the narratives of emergency and crisis propagated in Brazil during the second half of May 2018. Despite the different amplitude concerning scales, the rhetoric of securitisation encompasses the two discursive formations as events that generate an ontological insecurity. People feel helpless, sensing "a lack of agency" that leads them "to ignore the issue and stick to a false sense of security in their home and community" (Harries 2008 cited in Warner and Boas 2017: 212).

It is meaningful that the baker who sells at the organic farmers market associated the exaggerated movement of consumption, due to the fear, with a possible behaviour of people facing "the end of the world".

Danowski and Viveiros de Castro (2014) suggest there are many discourses on the end of the world. Nevertheless, every apocalyptic discourse shares the notion of a "general collapse of the temporal and spatial scales" (30). The idea of the end of the world evokes the dissolution of time and space. The discourses on climate change often present this terrifying possibility of the finiteness of human existence. In a sense, the discourses on the situation of emergency disseminated in those days of May were surrounded by the fearful possibility of the finiteness of food that supplied middle and high classes in the cities. If the lack of nourishment leads some people to create scary imagery and act catastrophically, as if they were living the end of the world, the population who live in the streets and do not have food know many worlds have already ended a long time ago.

There is also another relation between the alarmist rhetoric on climate change and the situation of emergency in Brazil: if the smokescreen of the numberless news and hasty reactions is dissipated, the underlying issue is the political use of these catastrophic discourses. According to Warner and Boas (2017: 206), "presenting threats as a catastrophe may even be (seen as) *instrumentalised* for ulterior, political (or moral) ends". In the case of the conjunction of political and economic crisis in Brazil, expressed by the strike of the truckers and the alarmist discourses of the end of May 2018, it is possible to observe that authoritarian political movements (which present themselves as "popular" and "apolitical") seized the unstable moment to manifest obscure and antidemocratic agendas. Such movements raise their voices for military intervention, which is a euphemism for dictatorship.

Nonetheless, the discourses on climate change – and on the environmental issues too – are used as a political strategy in electoral campaigns, but also as part of government projects that take advantage of "the level of social articulation on the theme that can be mobilised politically" (Ryan 2017: 282).

However, the political and economic disputes concerning fossil fuels are also discursive debates on the uses and abuses of "natural resources".

The issue of climate change involves more evidently the anthropogenic action as a destructive geological force (Danowski and Viveiros de Castro 2014), in part due to capitalist appropriation of the environment, understood as "natural resources" and commodities. Nevertheless, the issue of fossil fuels is also embraced by environmental policy in the context of the neoliberal order. The ideology of development that justifies the predatory exploitation of the environment is rooted in modernist thought as a model, "an evolutive, linear and unavoidable way that underdeveloped societies must follow in order to overcome poverty and retardation" (Scotto, Carvalho and Guimarães 2007: 16). Technological development, which is usually assimilated by capitalist regimes of accumulation, is thus considered the zenith of progress.[1]

Regardless of their positions on the political spectrum, different projects of government – by means of politics of scepticism, indifference or low-intensity integration of environmental issues (Ryan 2017) – tend to consider environmental issues as obstacles to development. In some contexts, "anything sounds more offensive in the public debate than the accusation of being against development" (Fleury and Almeida 2013: 152). That is the reason why there are different environmental conflicts related to "natural resources" and symbolic controversies regarding "sustainable development". Furthermore, complex social effects are inherent to the logic of expropriation of "Nature", which is often understood as a reality external to the human and as a source of material resources used to uplift national economies.

The current situation in Brazil is different from the context of economic growth established by big projects of development, such as the building of the Belo Monte hydroelectric plant a few years before (Fleury and Almeida 2013). The political and economic dispute regarding Petrobras involves contrary views on national sovereignty. Many governments prefer privatisation of Brazilian companies, exporting goods and commodities to so-called developed countries, whilst the Brazilian population pays high taxes buying the same goods from foreign companies. The Brazilian propaganda of petroleum self-sufficiency is such an example. The shattering of such a conviction renders some people afraid of an apocalypse.

Note

1 The urge for development can be understood as a kind of discourse embedded with modernist thought. I understand modernity as the concrete political project that became hegemonic in Euro-American societies and present practical consequences (see Asad 2003). Besides the political aspect of the discourses founded in the idea of modernity as an ideological construct, I understand there is a modernist modality of thought, which is based on specific onto-epistemic assumptions.

References

Danowski, D., and Viveiros de Castro, E. 2014. *Há mundos por vir? Ensaio sobre os medos e os fins*. Desterro [Florianópolis]: Cultura e Barbárie; Instituto Socioambiental.

Fleury, L., and Almeida, J. 2013. "A construção da usina hidrelétrica de belo monte: Conflito ambiental e o dilema do desenvolvimento". *Ambiente & Sociedade* 16(4): 141–156.

Ryan, D. 2017. "Política y cambio climático: Explorando la relación entre partidos políticos y la problemática climática en América Latina". *Ambiente & Sociedade* 20(3): 271–286.

Scotto, G., Carvalho, I., and Guimarães, L. B. 2007. *Desenvolvimento sustentável*. Petropólis, RJ: Vozes.

Warner, J., and Boas, I. 2017. "Securitisation of climate change: The risk of exaggeration". *Ambiente & Sociedade* 20: 203–224.

2 The way of the Muriqui-Assu

Figure 2.1 A permacultural house

On a long dirt road, in Niterói, there is a neighbourhood of small farms, situated in a buffer zone of Serra da Tiririca.[1] Bucolic, the neighbourhood of Muriqui is close to the south part of São Gonçalo, to which one could access by the road Estrada Velha de Maricá. The name of the region derives from a species of enormous monkeys (*Brachyteles arachnoides*) who inhabited the region decades ago, and are facing extinction today.

On the road Estrada do Muriqui Grande, also called Road Aristides de Melo, resides a small collective of people dedicated to Permaculture,

DOI: 10.4324/9781003378853-4

Bioconstruction and Agroforestry.[2] Well-integrated into the neighbourhood, they offer the vegetables, fruits and bread that come from their gardens and kitchen. The smallholding consists of two contiguous pieces of land, cut by the road. On the first land of 10 hectares, there is a house, the main headquarter of the Project of Ecovillage Muriqui Assu. Inside the house, there are the kitchen and rooms used to shelter the few dwellers and guests. Besides the house, there is also an enormous hennery, a garden, a dry toilet, a worm farm, many trees and plants, some of which were seedlings and others already rooted in big vases. In the other land of 5 hectares, the future headquarter, a kind of mother house, was still in the process of bioconstruction (Figure 2.1). There was also a small agroforestry system of symbiotic plants.

Reciprocity beyond the human

In August 2018, when I visited the Project of Ecovillage Muriqui Assu, Wilson and Priscila, 61 and 57 years old, respectively, coordinated educational activities by means of experiences in community, including courses and volunteering. The Project aimed, in the long run, to build more housing adjacent to the mother house.

Breakfast was always served collectively, at 7 a.m., after a ringing bell that awoke everybody. At the table were two young men, who were visiting the Project as volunteers: a 19-year-old Colombian, who studied territorial development, and a 24-year-old black person born in Rio de Janeiro, who studied accounting and taught surf. Over the table, there was home-made whole bread, butter, coffee, milk powder, brown sugar and scrambled eggs. Curiously, the conversation topic was manure. An inside joke. Wilson said we needed to maintain our faeces composting for a long time, putting sawdust and letting microorganisms work. This process would become good manure. Human excrement was left in the composter for six months. Like other unexpected materials, human excrement was fundamental to permaculture. The premise of not-losing-anything-in-nature is definitely followed. There are also, according to Wilson, microorganisms very similar to those coming from humans. They can be obtained by the cooking of rice that, after 15 days in the forest, becomes rotten and presents a bluish crust. Such a crust, created by microorganisms, is an excellent cleaning product because the microorganisms literally feed on dirt.

After the instructive conversation at the table, we were led to the structure which was being built according to the techniques of bioconstruction.

Our task was to withdraw sand from the ground, sift it and bring it together with the other materials. There were horse faeces and milk upon a plastic groundsheet, which was mixed with the sand and a little bit of lime. The procedure was trampling the cob to homogenise it. The cob is ready when it sticks to the hands, even if one turns the hands down. The cob would be engrafted alongside a longitudinal bamboo structure, vertically and horizontally, which

Figure 2.2 Nature provides the design

was the wall of the building. This was the basic technique of cob wall (pau-a-pique). The mother house was being built by many hands, especially the volunteers and attendees of the courses.

After lunch, I chose to help Nicolas with the plants. He was the son of Priscila and Wilson. He was 21 years old and always answered every question with gentleness and attention. He told me he lived in a community of the Family International, a Christian institution once called Children of God.[3] The other ten children of his father also grew up in communities. Nicolas was homeschooled until he was 12 years old. He told me that he had no religion in the conventional sense of the term, but he believed in the force of nature. He thought people who live in farms or close to nature are more sensitive to the turbulence of city life because the communities are not separate from the world and the social problems really affect the dwellers.

Nicolas introduced me to some principles of permaculture and agroforestry. First, human beings can be integrated into cycles of nature through simple practices like alimentation. The residues which are often considered garbage can be reused. The residues become manure to the plants, which later feed humans. Even the parts humans do not eat can become organic manure by means of composting or serve as nutrition to other animals. Nicolas and I were gathering the plant trapoeraba (*Tradescantia zebrina*)[4] to feed the hens. Nicolas told me that the ground of a healthy forest is always covered by many leaves. The leaves protect the soil from the weather, keep the earth humid and generate shadow for insects, such as ants and worms. An unprotected soil, without leaves, is dry.

Regarding the use of pesticides to prevent proliferation of insects in the garden plots, Nicolas understood that excessive ants and other species revealed some unbalance in the soil.

Even though he was very young, Nicolas held a vast knowledge connected to agroforestry and permacultural practices, due to his intense dedication to the Project, his daily experience with the earth and the courses he attended. This lived knowledge is fundamentally experiential, founded on the perception one has of the environment. It is not only about concepts, but rather about knowledge as a practice (Lave 2015).[5] And such a knowledge is revealed through engagement in the world, when the person inhabits the environment and knows it from within (Ingold 2002, 2005).

On the following day, August 14, 2018, a lady called Inês visited the Project for the first time. She integrated a popular library that enacted cultural resistance in a district of Niterói (BEN – Biblioteca Engenho do Mato). She was a permaculturalist who had some experience with her own land. She first helped us remove little plants off of the soil, leaving them as green manure on the plots. While Inês and Nicolas chatted, I was listening to them and withdrawing some leaves from the bottom of stalks of mustard and kale, in order to stimulate growth. Then, I cut branches of the castor bean plant and deposited them on the soil as organic material, which becomes nutritious manure after decomposition. In the agroforestry system, every plant was integrated to the whole homeostatic process. Nicolas and Inês shared their ideas and they understood each other very well. Nicolas said: "We want biodiversity". And Inês added: "For life". They did not understand insects as plague, because the insects and other species were indicative species – they brought the sign of some process happening on that soil. If there is too much clover on the ground, the clover indicates the excess acidity of the soil and, at the same time, helps to solve the problem. If any animal or insect eats the produce, it is a sign that the proper moment of harvesting was lost. Inês expressed that she did not mind sharing her harvest with other animals, when they came to eat it. They needed nourishment too. Her thought was only possible outside a logic of capitalist accumulation. Reciprocity was a core tenet of permaculture philosophy, including humans and non-humans.

Redistribution of surplus, earth care and people care are the ethical principles of permaculture. The practice of permaculture has its own dynamics, based on flowcharts that represent cycles. Permaculturalists always talk about closing cycles. If a dam, for example, provides fish, the residues of fish return to the compost, which become manure for the plants; the plants, in turn, return to the house, feeding humans and so forth. These flows of materials among humans and non-human communities – vegetables and animals included – can be easily followed and people know from where and to where each element comes and goes. Even though the human perspective seems prominent in some discourses of my interlocutors, permaculture can be understood as an ecological epistemology that decentres the human and presents a symmetry between the human world and the more-than-human world (Carvalho 2016). The relation between humans and non-humans, through the entangled flows, can only be symmetrical in an "environment without objects" (Ingold 2010). The environment in which plants, excrement, residues, organic matter and other things are entangled and move along vital lines as co-creators of reciprocity cycles beyond the human.

Imagining the culture of the forest

At dusk, on one of the ten days I was there, I could interview Nicolas formally. We had had the opportunity of visiting the agroforestry that Nicolas was cultivating in the small glebe. Despite being small in scale, as compared to the big agroforestry cultivated by Ernst Götsch,[6] the agroforestry Nicolas was taking care of brought him some experience.

We sat on the ground and an enormous black dog called Argus approached us, sniffing our feet. Nicolas taught me:

– The idea of agroforestry is to create an environment of interaction and not exclusion of the human being, unlike areas of permanent protection or forest reserves. The proposal of Ernst is creating an area of permanent interaction. This is done from the beginning because the human plants a tree and next to that tree a wide range of other things which will be harvested in different times. So, you will not reforest the place, putting trees and returning there no more. You will return every week because you will soon harvest the radish – which is the first thing you can harvest, really quick – and then the lettuce, after one month or so.

Listening curiously and attentively about how the process of an agroforestry begins, I asked: "And the arugula?" Nicolas answered:

– A month. Pumpkin, four months. Manioc, one year. And the system is renewed when you prune the already growing trees. In one year, after the first harvest, banana trees would be grown too and then you must manage more intensively. It is possible to prune some trees or open a new glade. The renovation sends the information of renewal to the whole area. If you cut your hair, growth is stimulated. The plant also. If you prune above, then the roots will die below. If you deposit manure on the soil, it will become more fertile due to the rotten organic matter. So the renovation of the area stimulates the new life which will come there. Every tree will be sprouting and you can plant a bunch of new seeds again. And all life follows. This will begin a new cycle. This is one form of agroforestry. There are many forms. We are planting our system in nests and the management is more or less like that. I think it will always be more or less like that because the agroforestry systems normally are formed by two plats of trees or tall plants. And between them, three or two plats of vegetables. There are vegetables on the glade and fruit trees on the lateral plats. The trees of these lateral borders become very tall and shadow the vegetables. So you can begin the planting of more shadow trees. Coffee and cocoa like areas with shadow. Together, everything becomes a forest. Some trees are well established. Some trees need intense pruning because they feed the soil. The idea is planting the food to each plant. The book *Agroflorestando o mundo: do facão ao trator*[7] explains all these relations. It is very important to know how nature works, how this organism Earth works. A self-regulating organism. This is Gaia theory, isn'it?

Yes – I said – this is the concept of homeostasis, the ability of each organism in maintaining its own life. And I added: – Let me ask you another question on this idea of agroforesting the world. On a global scale, there is this hegemonic monoculture in the capitalist regime of accumulation…

Nicolas responded: – … which follows the concept of scarcity, always…

While a coconut fell from its tree, I added: – There is scarcity and abundance…

And Nicolas responded: – The law of abundance and the law of scarcity. Society always revolves around the law of scarcity. The monoculture with pesticides and herbicides came with the green revolution and the remaining war machinery that came to the countryside. Barbed wire, for example, comes from the war.

– On one hand, we have monoculture with capitalist purposes and, on the other hand, we have agroforestry and other similar techniques…

– … which can also be used for capitalist purposes! – asserted Nicolas.

– Would you say it is possible to expand agroforest in a more universalised way? In the sense of making this kind of cultivation grow and gradually or quickly replace the other? Do you think agroforestry is feasible?

– Of course! There is much more abundance. On the same land people in monoculture only plant lettuce, I can plant lettuce and two other things around it. I can seed beetroot and kale at the same time. I would be producing the double. Automatically, one initial system of agroforestry has much more abundance than one monoculture system. Can you imagine the trees growing in that system? Fruits, real food for human beings. Lettuce is not food for animals like us. We would need lots of lettuce if we were to be fed only by this! Humans are forest beasts and need big fruits and prey. The forest brings that. It brings animals. The forest planted by Ernst is one of the most biodiverse forests in Brazil. Maybe the most biodiverse rainforest. Diversity brings balance. More biodiversity brings more balance. More food to all insects and more nutrients to the soil because each plant brings nutrients to the soil. Biodiversity begins with small animals, which eat lettuce, little seeds, grass. This is the first stage. Later, larger animals that eat the small ones. In the forest planted by Ernst, there are gigantic snakes that feed on poisonous bugs. There are also monkeys and onças. The region was a desert when he began to plant. And this is what I mean when I talk about changing the culture. Our society eats things, but does not feed on the forest. People do not feed on the forest and do not live in communion with the forest. But the forest is the way of abundance.

– And do you think replacing the hegemonic model of food production is possible?

– Totally, the young man said. Evidently, he was not considering how the capitalist system expands its development ideologies or the complex political process by which capitalism intervene in racially and economically subordinate communities and societies, articulating them to the global market (Quintero 2015).

– How? – I asked, with a hopeful tone of voice, noticing that permaculture and agroforestry are resilient life projects of extractive communities that escape from commoditisation and the accumulation logic (Zhouri, Laschefski and Pereira 2005).

– It depends on society supporting the agroforestry people to create even more. We need big farms to produce in an agroforestry system. There are many. Many families already live in agroforestry settlements.

Later, reflecting on the economic and geopolitical aspects of the topic, I understood the example of families in settlements as a form of resistance and as "response of local communities to the development interventions" (Quintero 2015: 71). Our conversation reminded me of disputes for environmental justice, particularly the conflicts on "ownership, management and control of the space", where "the uses of dominant segments of society, which impact the territories and resources occupied and mobilised by vulnerable groups" (Zhouri and Oliveira 2010: 448). I remarked: – There are many people who do not own lands. And I am not even talking about the movement of rural workers (MTST). Most people do not own lands, they do not have any property, a piece of ground, right? But the big landholders do have those enormous terrains in which they cultivate…

– … big monocultures of soy, corn, Nicolas amended.

– … big monocultures of soy and sugarcane, and the like. Thus, the scale of land, the scale of power, the amount of land and power those hold are very different from the power and land of the agroforestry practitioners. Do you understand the difficulty I am trying to show here?

– Yes and not much. I don't really agree, Nicolas said, surprisingly.

– Why?

– Because there are already landholders who are adopting agroforestry. What does capitalism want? Money. If you prove they can make double the money, and also retire from working with wood, what more will those people want? Capitalism wants money and I think this is something happening little by little. It is growing in Brazil. I do not know what it is like abroad. Fernando, who instructed me in agroforestry and participates in courses here with us, is going to Portugal, where he will offer two courses of agroforestry. It is expanding and spreading. And permaculture has also been inspired by biodynamic agriculture, by the straw revolution, which is planting upon straw. Fukuoka, a Japanese farmer, developed this. And permaculture came as a response to the green revolution, for a more permanent agriculture… Permaculture.

– Is permaculture a response to the green revolution? I asked, surprised.

– Exactly. In the beginning, it was the permanent agriculture, and then it unfolded to a permanent culture. This is already big, taking care of the soil, taking care of the earth, but capitalism does not release this information easily. Capitalism does not want permaculture to be divulgated because misery and poverty are widely a part of the capitalist process. It is the abundance of scarcity. But there are long-lasting successful experiences. Many families noticed that permanent agriculture is abundant and saw the fruits. The people who followed the traditional system say: "I do not want anything but this. This is everything I needed. My income tripled". I have heard many histories. So, it is possible. But there is the issue of the forest culture. People are afraid of the forest. The fear of the dark. The forest seems dark. People fear the beast… the danger… This is part of our culture and needs to be changed. Agroforestry is changing the culture. I think everything is related to everything… all the same scheme… the witches and everything that religion suppressed and did the same with the forest, with nature, creating paradigms.

This digression from Nicolas was reasonable. According to the historian Starhawk (2018), the suppression of supposed witchcraft was a lasting movement orchestrated by Christian institutions that tried to facilitate the transition from feudal systems to capitalism. Thus, "the persecutions of Witches arose as a manner of contesting and eradicating the ancient perception that the world as alive and life has a perception and consciousness" (Starhawk 2018: 57).

The process of suppression of experiences of the ancient Euro-American populations who lived in connection with nature is intrinsic to the passage from an animist ontology to an anthropocentric ideology, which has considered nature a non-living thing, an object to be explored by the technocratic development. The fear of the forest, mentioned by Nicolas, also expresses the fear of an unknown force, the very life that pulsates in the natural elements. On the other hand, it expresses the separation between society and nature, between the unconscious non-human and the human who constructs culture. Such a disjunction of culture and nature "has functioned all these years as a cosmology conforming our consciousness and the political organization of the institutions that set modern society" (Steil and Carvalho 2014: 170). Nevertheless, these are only some aspects of the historical process that led to the current ecological crisis.

Sharing as a mode of life

Every day, by morning, we watered the nursery of seedlings. One of the proposed tasks was its organisation. Marcos and I wrote the names of each species on stickers, affixing them to the little black bags in which rested the little plants. Some seedlings, bigger, were planted in vases. There were also wooden boxes with little bags of the same species. We classified the similar species, moving and grouping them, so they would be close to each other (avocado with avocado, citrus with citrus, cupuaçu with cupuaçu, açaí with açaí and so forth). We also counted the quantity of fruit trees, taking note of each size. It took many days to identify, group and classify the plants. Sometimes identifying the plant only by the eye is not easy. The structure of the leaves indicates the species, but subtle differences cause confusion. For this reason, Marcos and I obtained support from Fernando, who was visiting the place one afternoon. He was able to easily identify some species that nobody knew. There were 308 seedlings of at least 25 species in different stages of growth. Some germinating; others more than a metre tall, urgently needing to be transplanted to the soil. Most of them were cocoa trees.

Near the nursery, there was a dry toilet, bioconstructed according to the technique of adobe. This kind of toilet does not need water. The human waste was always covered with sawdust, and then, after six months in the compost, it would become excellent manure that, despite not being applied to vegetables, could bring nutrients to big trees and fruit trees.

Close to the dry toilet, there was a worm farm. The worm farm was another important source of manure. Organic residues were frequently deposited to feed the worms. There was sawdust, organic matter, leaves and chicken excrement. These materials were worked by the worms, generating a highly effective fertiliser for the garden plots. It was important to always keep the worm farm humid, covered by tarpaulin. For this very reason, we always seized the watering of the seedlings to moisten the worm farm.

Priscila watered the plants frequently. She was the matriarch of the Project. Always active, she took care of the cleaning and storage of eggs in plastic trays, which were sold to the neighbours. She also cooked, to guarantee the abundance of groceries. She organised the delivery logistics of organic product hampers, and she also made bread and other foodstuff.

With her mild smile, Priscila invited me into a conversation with the neighbourhood of Muriqui. The topic was a misused collective dump at the road Estrada do Muriqui Grande. The problem was the starting point of an exposition on permaculture, whose aim was to raise awareness about the necessity of recycling in the community. Priscila and Wilson had submitted a proposal that received a small financial support by the programme Active Citizens.[8] The programme supported actions of sustainability in developing countries. Priscila and Nicolas had participated in a short-term course of leadership, offered to women and youth.

The invitation was made in advance and Priscila tried to reach all the 300 people who lived in the region. She also prepared snacks and I helped her with the passion fruit juice, on that amenable Saturday afternoon.

Across the little market, which was also a bar, there was a soccer field. In this community centre of Muriqui Neighbourhood, we set a space with plastic tables borrowed from the little market. Upon the tables, the snacks. Around the table, chairs. Attending the meeting, 20 people.

Lúcio, actor and movie maker, who was a guest of the Project, introduced some concepts regarding organic gardens, and later distributed seed to everybody. They listened, their eyes bright. A collective garden, cultivated by the neighbours, could be doubly useful by solving the recycling problem and providing food. The collective dump in front of the community centre was built

Figure 2.3 Awakening the neighbourhood

by the Association of Dwellers of Chibante Street, a street that was crossed by the road Estrada do Muriqui Grande. The trash was taken by the public garbage trucks. Nevertheless, the dump was only a part of the problem. Next to the concrete dump a "stream of waste", as Priscila called it, was formed. This flow of water, which sprung from a smallholding nearby, was now almost stagnant and fetid. Its boundaries were full of plastic residues, pieces of furniture and electronics. The stream flowed to the opposite side, in relation to the place of Priscila and Wilson, heading to Chibante Street and Patience Street.

"Many springs in the region dried", we heard Priscila say, exasperated. And a dozen boys played soccer on the court inside the community centre where the meeting was taking place. Sometimes a single whoop crossed the space. Other times, voices rose simultaneously. The conclusion was that there was a collective interest in building a point of selective garbage collection there, with ten instructional signs. A follow-up meeting would take place in a month. The objectives would be recycling oil, organising the dumpster, initiating an organic garden and donating organic residues to the Project Muriqui Assu, which would compost them. At snack time, many people appeared. The boys approached the table and heard some words regarding recycling.

Days before, when I had explained my research to Priscila, she had told me she agreed with my understanding of spirituality and ecological thought as two modalities of connection with the world, the earth, people, with other living beings, with the universe and so on. She had asked me something, and I was left thinking about it. Would it be possible to experience spirituality in the cities, where people are immersed in a voracious system? We decided to record our dialogue after returning from our meeting with the neighbours. We were able to deepen that question and observe such a perspective, rooted in the lived experience of Priscila.

We sat at the big table on the porch. Priscila looked at me attentively, with her peculiar, welcoming glance. She told me she lived in other communities for more than 15 years, from 1993 to 2009. The first community she lived in was Christian. She met Wilson there. They became partners and Nicolas was born. In this community, there were 25 people. By 2009, the youth were finishing high school and wanted further studies. Thus, for this and other reasons, the community understood the moment required care for the young people who were no longer interested in being part of the Christian group. Big communities dissolved and the couple came to live near Niterói city centre, which was still a region considered rural. Sometimes, friends appeared asking: "Oh could we stay here with you for a period?" When, finally, Priscila and Wilson decided to live in the smallholding, focusing their efforts on permaculture and bioconstruction, they opened their house to guests because they thought that only their family would be monotonous. When I visited the Project Muriqui Assu, the house was receiving ten people, among them volunteers, supporters and people interested in integrating into the community life.

Bearing in mind the conjunction of emotion, action and reflection as the fundament of the lived experience, as Turner (1986) taught, I asked Priscila: – How did life in community look like to you? I know the community was part of The Family International, which was before known as Children of God. I would like you to talk about this experience you lived, your reflections. It will surely help me to understand the concepts, the conceptions of the community.

Priscila expressed the reasons why she decided to join the community and how this happened: – Well, when I joined the group, it was not Children of God anymore. When the offspring grew up, the focus of the community changed. In the beginning, everybody was unattached ex-addicted young people. Around 1978, 1980, the shift to a family setting began. The issue was school and formation of the children. Before joining The Family, even when I worked in the traditional system, I had always wanted to live something different. I disagreed with how things go. I was even affiliated to the Communist Party; I thought I could change the world, but I saw it was not the way. Communism had some points, but it lacked love to care for people and nature. I was always unsatisfied with the traditional way of life. When I discovered the community,

I had been a lawyer for ten years. I was successful in terms of income, house, car, relationship. But I really felt anguished. Because I apparently had everything I wanted, but I did not have peace of spirit. One day, I went to this bar, a restaurant. And a guy was singing the songs of the missionary group Children of God, which did not talk about God explicitly. For example, one that touched me deep said: I will go on singing. I will go on loving. I will go on distributing life. Wherever I go". He sang many other songs. Argentinian music and Popular Brazilian Music. And between the songs he played one that went like this: "Give me everything in life. Give me everything money cannot buy". The spirit I saw in him! I said to myself this was what I was looking for. I knew he had the message I wanted. When I went to meet the group, there were around forty people living together. People from different countries. I was enchanted by this way of life. I then started to frequent the community with a friend, for one year. And it was awesome. All those years, it was always awesome. I never regretted it, I never wished to return to that hegemonic way of life.

The Family International is an institution formed by communities of many kinds. The communities shared some philosophical Christian principles, but each community was free to create its own statute, according to the country and people who live there. There was certain flexibility, except for drugs and excessive drinking. The practice of evangelisation was the main goal for all communities. It was the golden rule. Priscila called it "pour on people".

The Christian message was shared differently, depending on the context of each community. In Brazil, for example, religious freedom is a right guaranteed by the state, notwithstanding the fact that Christianism is more accepted socially (Oro 2011, 2012). Even though the Brazilian institutions are considered secular – a very controversial issue (Montero 2015) – Christian religions have more declared adherents, expressivity in public space and, if compared to the minority religions, do not suffer from religious intolerance. In sum, Brazil is a territory where people can talk about Jesus anywhere. This allowed the Family International to evangelise, facing fewer barriers to come into schools or prisons, despite their communities being affiliated to a counter-hegemonic institution. In China, the missionaries shared their messages through English classes. Priscila did not tell me in detail how that happened, but it was obvious that various forms of evangelisation were used by the keen missionaries.

The operation of the missionaries was similar to the process of foundation of churches through the transnational webs of Gospel Brazilian institutions (Oro 2019). Even though Children of God can be considered a new sectarian religious movement, whose main idea was the propagation of Christian messages, the formation of communities followed a model of exclusivity and autonomy. The libertarian way of life in communities prevented the dialogue with other religions in the places where the missionaries passed. On the other hand, each Children of God community (also known as Family International) was autonomous in relation to each other, even if they were always in close contact through webs. People constantly circulated and crossed national frontiers to contribute to the formation and consolidation of communities or distribution of messages. Wilson, Priscila's husband, travelled to many countries, as well as some of his 11 children, who had been in Russia, Africa and other European countries.

We could hear voices chatting in the kitchen, while I asked: – What was your idea of mission?

– Sharing love, the message left by Jesus. The love of God for people. Independent of race, colour, creed, religion. We did not go to churches. But we believed and believe until today that the love of God can change lives. And we believe in the philosophy of Jesus. The Christian philosophy of forgiving, tolerating, loving the enemies. We believe in this was the main message to be spread. How? Through puppet theatre, distribution of pamphlets. Through distribution of videos, musical groups. Each missionary home had its own form of delivering the message. The message was, more or less, the message of Jesus in the New Testament.

Even though each community had autonomy to create its own statute, there were general guidelines that set the concepts and practices of the whole movement. The missionary practice, the core of the movement, aimed to spread the message of Jesus Christ and adapted to the reality of each region. In the communities, according to Priscila, there was a weekly alternation of activities, so that everyone participated in the communal life and the external life as well. People shared food and collectively paid bills. Priscila emphasised that "if something remained, it remained to everybody. If nothing remained, it did not remain to anyone". Notwithstanding the modest and harmonious life, there was conflict, which was a topic of discussion during weekly meetings that her community called "Home Council".

The horizontality of relations, through the equality of goods and creation of democratic spaces of decision-making, characterises a countercultural ideal, which is present in the history of alternative communities (Caravita 2012) and it is still manifested nowadays, for example, in the Rainbow communities and meetings (Cykman 2019). Indeed, other countercultural conceptions can be observed, such as the critique of the status quo and the search for alternative ways of living, often communitarian, which generates a tense interaction with the broader society. Such an issue can be exemplified by the necessity of dissolving communities when the children grew up and chose to live a more conventional life, within the "system". This internal conflictual process, involving the flow of members outwards, happened not only in the community in which Priscila lived, but in many other communities around the world. On the following day, when I interviewed Wilson, he – who dedicated a large portion of his life to the missionary movement – observed the same point. During our dialogue, he confirmed the challenge of the adults who tried to keep their children interested in the continuity of the community life.

Changing the topic, I wanted to know if Priscila remembered any spiritual experience she had living close to plants, animals and woody places, far from the city rumble.

Again, she answered in detail: – Nature. Walking. I like very much to hike alone because it is the time I feel communion with God, where I feel his presence stronger than I fell in the daily life, in the city. I have not lived in cities for many years, but in nature, of course, I feel more of the presence of God. The Bougainvillea that pours its flowers is the most beautiful thing. I feel the presence of God intensely, much stronger, and I can communicate with him all the time I am walking, especially alone. Walking in nature I feel closer to God. I feel him talking to me. I can only chat with God and cry talking to him when I am in nature. When I am hiking on the mountains, I can sense this deep connection. Sometimes, I start singing or manifesting out loud. But mostly it happens in silence. The rays of the sun on a cold day… I feel loved, embraced by God, by the universe, by this divine force, it does not matter how one calls it. I feel taken in the arms of a highest being. That warming sun on a chilly day warms within. This brings me the desire of thanking and, when I come in this spiritual universe, then everything can happen.

The spirituality lived by Priscila is intimately related to the experience of walking in nature. Even though her spiritual experiences happen especially during her lonely walking, such an experience is daily and loses its extraordinary character. It is rather the eruption of the sacred through the sensory experience of a divine force that makes itself present in the natural elements. During the hike, the spiritual experience generates a moment of contemplation that potentially harmonises "the bodily and spiritual state" of the person, in their relations with "a landscape that integrates, in a whole, humans and non-humans" (Steil and Toniol 2011: 46). The religious language expressed by Priscila – religious in the etymological sense of the term religare, which means reconnect with the sacred – unveils the perception of the immanent sacred of nature, which is founded by an "immanent spirituality in which the connection with the sacred happens through the elements of nature" (Comunello and Carvalho 2015: 97).

The life of a project

In 1976, Wilson experienced an extraordinarily memorable vision of Jesus Christ with open arms, saying: "Follow me". He was 17 years old. According to his narrative, such an experience was the beginning of his spiritual life in alternative communities. Moreover, it was the beginning of the journey that led him to permaculture and bioconstruction.

The next day, Wilson left his house with his Afro-textured hair and his ragged jeans. He went to a Catholic church, Igreja do Loreto, on Freguesia Street, in Jacarepaguá, Rio de Janeiro. He said to the priest: "Sir Priest, I am Christian since last night and I want to be baptised because my parents are atheists and I have never been baptised. What should I do?" Wary, the priest: "All right, son. Okay. You must come to church every Sunday, attend the service and, during the weekend, attend a course for you to understand what baptism is". And Wilson followed the instructions.

Wilson quit drugs, lost his former friends and began to experience a personal communication with God. Even through his intense frequency at the church, the young Wilson noticed that the Jesus Christ people talked about was different from the one he met, during his trance, that night. As many other people in New Age contexts reported, Wilson perceived he had an intimate relationship with God, a "strictly personal" conversation with the sacred (De La Torre 2011). His experience is related to the decline of religious traditions and the massive experience of inner-life spiritualities (Heelas 2006). Wilson was unsatisfied with that difference he noticed and, in other church, he prayed: "Jesus, I believe you are real, you have already manifested yourself in my life. But now I do not want to live in this world anymore. I want you to take me from this world or I want to live just for you". Curiously, a few days later, he met a person who gave him the key to unlock his difficulty in associating his spiritual experience and daily mundane life.

Inside a crowded bus, a Jamaican man was delivering pamphlets and spreading a Christian message. He was an affiliate to the organisation Children of God. Something in the smile of the missionary bothered Wilson. The love Wilson felt emanating from the men bothered him. "No, I do not want your message". The missionary insisted. And mentioned the well-known band Meninos de Deus. And he also mentioned the alternative community. Wilson knew about the existence of a community in Grajaú, inhabited by some people he considered very wacky. "Come visit us, we will have a party this Saturday", the Jamaican missionary said. Trying to get rid of him, Wilson answered affirmatively. And took the pamphlet.

On Saturday, Wilson remembered the invitation, and felt a willingness to go to the place. He did not have the address and, undecided, he took a bus. He went, asking for information from the people who lived in the neighbourhood. And he found it. When he arrived there, he was impressed by the spirituality of those people. He was received with Bible verses, and he felt loved. He felt it was his home. He thought that such a way of living was the closest to what Jesus taught to his disciples. And he decided to join them.

For 35 years, Wilson dedicated his life to the organisation Children of God, which later became the Family International. He lived in 30 different cities, and in other countries. He was a pioneer and helped to build many communities. He was a house builder, literally. His skills in construction were developed. Wilson followed some ecological principles intuitively, such as septic tanks and rainwater harvesting.

Nevertheless, when the children grew up, a conflict surged in the hundreds of communities worldwide. Many young adults did not want to live the life their parents chose. Wilson also noticed this conflictual process in communities bound to other organisations. The dissolution of communities gave rise to the problem of money. The financial management of the communities of Children of God was different from other common Christian institutions. Generally, the proselytes did not have money and the income came from donations or temporary work. In some cases, the dedication of the missionaries approached mendicancy (Beckford 2010). Wilson had never worked in the "system" and felt worried about the subsistence of his 11 children. However, he was able to use his skills in typography and printing, acquired throughout the years he was responsible for the small printing office of the group, in which they produced pamphlets and a monthly magazine. Wilson, quitting community life, opened a print shop and began to print business calendars with Christian messages. This was the form he found to maintain his mission.

Years later, Priscila and Wilson were dispossessed of their land in Serra da Tiririca. The land became part of the Public Park Serra da Tiririca. They later bought the lands in which they lived when I met them. Their initial plan was to build sustainable houses. Wilson started his studies on sustainability, with commercial purposes. Nevertheless, he discovered bioconstruction and permaculture and created the Project of Ecovillage Muriqui Assu. He participated in many activities on permaculture and completed his Permaculture Design Course (PDC), a course in which people experience intense community life in an ecovillage for two weeks. The participant learns the ethical and practical principles of permacultural design and receives a certificate.[9]

Wilson noticed that permaculture promoted alternative communities. For his many years of community life, he identified some points of contact between the principles of permaculture and the conceptions and experiences he learnt alongside his Christian life. The idea of loving others and sharing, to his mind, was similar to the permaculture ethics: care for the earth, care for the people and sharing of the surplus. The flower of permaculture contains many different aspects within its petals, and one petal represents spirituality, particularly in its association to health and well-being. Such a notion of spirituality is diluted in daily life (Holloway 2003). Except for a little framed image of a white Jesus Christ in the living room, there was nothing particularly religious or openly spiritual in the smallholding where Wilson and Priscila lived. The spirituality they experienced were the very ethics of permaculture, assimilated with the conception of "love as religion" which, according to Wilson, was the practice of "making people feel loved".

When I met Wilson, who was then 61, he told me that spirituality meant care and love, including for the earth, not only the humans among them. His equation of permaculture ideas with his previous experience in Christian communities was at the ethical level. It was also the base of a mode of living that could be considered a kind of ecological asceticism in the immanent world. Instead of the intramundane asceticism of Protestantism (Nobre 2006) – in which the success within the capitalist structure of modern society would be the sign of the spiritual development of elects (Weber 2005) – Wilson experienced the refusal of the current system, as well as the distancing from Puritan asceticism. Wilson's equation points to an eco-religious conception that is inconsistent with some Biblical assumptions, especially the one according to which humans are separate from other beings (Giner and Tábara 1999).

Wilson's current experience was mainly influenced by his previous spiritual life. When he told me about the relations of communities and the broad societies, Wilson said: "We lived for the world, for the people in the world". The meaning of the intramundane social action was the diffusion of the Christian message. When he left Children of God, Wilson began to spread the message of permaculture. This is the main aim of the Project Muriqui Assu. His lived experience conjoined Christian ethics with the permacultural principles and led him to experience his mission within communities, not isolated from the surrounding social problems.

Permaculture, whose objective is to propose a permanent culture for society, is based on three principles and on a set of other premises. It forecasts an alternative mode for social relations. Its aims are fundamentally transformative. Even though Wilson avoided the term "political", in the conceptual framework of permaculture, the transformation of living is not circumscribed to familiar units or small communities. The spiral contained by the Permaculture Flower points exactly to this expansion of permacultural movement, from the personal level to the local and from the collective to the global. And this countercultural character permeates the creation of alternative communities that galvanise the youth unsatisfied with the capitalist culture of consumption and accumulation – and provides them with a sense of belonging. In addition to the critiques of hegemonic values, permaculture proposed and proposes "a message of hope in the struggles against environmental and social evils", facilitating "the experimentation and pioneering of lifestyle models directed by the ecological imperative" (Holmgren 2011: xxii).

Again, Wilson's experience and his permacultural project are rooted in the wide preoccupation with the future of coming generations which will live in the planet. His engagement is diffusing permaculture as a possible way of life in which humans are in communion among themselves, with nature, God and the universe. This ecological rationality is properly an ethics that includes a set of techniques and concepts for human adaptation and resilience in the face of the social and environmental catastrophe. The optimistic perspective of Wilson partially loses sight of the political implications necessary to the social transformations that permaculture proposes. It is an expression of "cosmic piety" whose aims and final consequences are the "continuation of the flourishing of human life in the framework of a better society" (Giner and Tábara 1999: 75).

The spirituality lived by Wilson had always been mystic in the sense of emphasising subjectivity and personal experience. Only in his mature years did he become interested in an ecological spirituality associated with this individualistic mysticism. His life crossed diverse possible modalities of accessing the sacred in a secular context of individual autonomy that favours the free choice of the subject (Frigerio 2016). The particularity of his experience is that, notwithstanding his passage through the church, the sect and the mysticism (Toennies et al. 1973), all the impulse of his trajectory was directly tied to his memorable spiritual experience as a young man: the vision of Jesus Christ who called for him.

The evening was setting down the porch where Wilson and I talked. Twilight left us. Floating on the swinging hammock, Wilson said: – The idea of permaculture is not only caring for your own house, but also for your city. It includes the whole governmental system. But if it exceeds 50 thousand people, it is not possible to apply any model. Permaculture teaches how to take care of the soil, the water; how to take care of the agriculture and the forest. Thus, it tries to integrate everything while observing the best of the ancestral cultures and of the available technology. They are brought together in the proposal of an ideal governance. But I am sure that man is selfish in himself. If the man does not search for his own spirituality, the growth of his love, he is, by nature, selfish. Thus, no model of governance will work if people do not change within. I believe this personal search is necessary.

Notes

1 Buffer zones are "areas peripheral to a specific protected area, where restrictions on resource use and special development measures are undertaken in order to enhance the conservation value of the protected area". Available in https://www.biodiversitya-z.org/content/buffer-zones.
2 Alongside the following lines, concepts and practices of each will be described in detail.
3 The Family International, originally called Children of God, is considered a new religious movement by the sociologists of religion (Beckford 2010). It was founded in 1968, in Los Angeles, by the former protestant reverend David Brandt Berg. The diffusion of the movement reached more than 50 countries and instituted at least 200 communities (Rodrigues 2008). The movement stems from counterculture, subscribing to some hippie ideas, such as sexual liberation and free love. Its missionary practice is considered "flirty fishing" because, for a certain period of time, missionaries used sex as a form of proselytism, through which they aimed "to demonstrate the love of God and obtain conversions" (Jungblut and Adami 2017: 105). Years ago, there were denunciations concerning sexual abuse on the part of the spiritual leader, including children.

4 A non-conventional food plant, also known as ondas-do-mar in other regions of Brazil, is a climbing plant that usually sticks to the ground. Its leaves are purplish and the taste sounds like cabbage.
5 Knowledge as practice, according to Lave (2015), regards the cultural processes through which people learn techniques and crafts. The forms of knowledge are entangled with social life, and changes in the practices of apprenticeship correspond to social changes.
6 Swiss agriculturalist and researcher who lives in Brazil. He developed a set of principles and techniques known as Syntropic Agriculture. Internationally recognised, Götsch is considered a true hero of ecology. In the eighties, he bought the Farm Olhos D'Água, 480 hectares, in the village of Piraí do Norte, na Bahia. The land was formerly destroyed by deforestation and, with the practices of cultivation created by him, even the springs resurged. More information can be found at https://www.renature.co/articles/ernst-gotsch-planting-water/.
7 Agroforesting the world: from the machete to the tractor.
8 The British Council leadership programme Active Citizens offered opportunities to the people and organisations engaged with social responsibility in contexts of sustainable development. According to the website, the programme "aims at the promotion of intercultural dialogue, social development and sustainable changes led by communities". Available in https://www.britishcouncil.org.br/atividades/sociedade/active-citizens
9 More information can be found at https://permacultureprinciples.com

References

Beckford, J. A. 2010. "Noveaux mouvements religieux". In *Dictionnaire des faits religieux*, eds. R. Azria and D. Hervieu-Léger. Paris: Quadrige/PUF, pp. 808–816.

Caravita, R. 2012, *"Somos todos um": Vida e imanência no movimento comunitário alternativo.* Campinas: Universidade Estadual de Campinas.

Carvalho, I. C. M. 2016. "Ecological epistemology (EE)". In *Encyclopedia of Latin American religions*, ed. H. P. Gooren. Switzerland: Springer International Publishing, pp. 1–3.

Comunello, L., and Carvalho, I. 2015. "Ecovilas: Aprendizagens, espiritualidade e ecologia". *Avá Revista de Antropología* 27: 81–99.

Cykman, N. 2019. *Limites do horizonte: Cartografia de uma episteme Utópica em encontros rainbow.* Florianópolis: Universidade Federal de Santa Catarina.

De La Torre, R. 2011. "Les rendez-vous manqués de l'anthropologie et du chamanisme". *Archives de Sciences Sociales des Religions* 153: 145–158.

Frigerio, A. 2016. "La 'nueva'? espiritualidade: Ontologia, epistemologia y sociologia de um concepto controvertido". *Ciencias Sociales y Religión/Ciências Sociais e Religião* 18(24): 209–231.

Giner, S., and Tábara, D. 1999. "Cosmic piety and ecological rationality". *International Sociology* 14(1): 59–82.

Heelas, P. 2006. "Challenging secularization theory: The growth of "new age", spiritualities of life". *The Hedgehog Review* 8(1–2): 46–59.

Holloway, J. 2003. "Make-believe: Spiritual practice, embodiment, and sacred space". *Environment and Planning A* 35(11): 1961–1974.

Holmgren, D. 2011. *Permaculture: Principles & pathways beyond sustainability.* Hampshire: Permanent Publications.

Ingold, T. 2002. *The perception of the environment.* London and New York: Routledge.

——— 2005. "Human worlds are culturally constructed". In *Key debates in anthropology*, ed. Tim Ingold. London: Routledge, pp. 112–118.

——— 2010. "Bringing things to life: Creative entanglements in a world of materials". *Realities Working Papers* 15: 1–14.

Jungblut, A., and Adami, V. H. 2017. "Hinduísmos ocidentalizados e suas percepções acerca do sexo: Movimento Hare Krishna e movimento Rajnenesh". *Religião e Sociedade* 37(1): 104–121.

Lave, J. 2015. "Aprendizagem como/na prática". *Horizontes Antropológicos* 21: 37–47.

Montero, P. 2015. *Religiões e controvérsias públicas*. São Paulo: Terceiro Nome/Ed. Unicamp.

Nobre, R. 2006. "Weber e a influência do protestantismo na configuração da modernidade ocidental". *Revista Cronos* 7(2): 289–301.

Oro, A. P. 2011. "A laicidade no Brasil e no Ocidente: Algumas considerações". *Civitas* 11(2): 221–237.

——— 2012. "Liberdade religiosa no Brasil: As percepções dos atores sociais". In *A religião no espaço público: Atores e objetos*, eds. A. Oro, C. Steil, R. Cipriani and E. Giumbelli. São Paulo: Terceiro Nome, pp. 181–193.

——— 2019. "Transnacionalização evangélica brasileira para a Europa: Significados, tipologia e acomodações". *Etnográfica* 23: 5–25.

Quintero, P. 2015. *Antropología del desarrollo: Perspectivas latinoamericanas*. Buenos Aires: Kula Ediciones.

Rodrigues, D. 2008. "Novos movimentos religiosos: Realidade e perspectiva sociológica". *Anthropológicas* 19(1): 17–42.

Starhawk. 2018. "Magia, visão e ação". *Revista do Instituto de Estudos Brasileiros* 69: 52–65.

Steil, C., and Carvalho, I. 2014. "Epistemologias ecológicas: Delimitando um conceito". *Mana* 20: 163–183.

Steil, C., and Toniol, R. 2011. "Ecologia, corpo e espiritualidade: Uma etnografia das experiências de caminhada ecológica em um grupo de ecoturistas". *Caderno CRH* 24: 29–49.

Toennies, F., Simmel, G., Troeltsch, E., and Weber, M. 1973. "Max Weber on church, sect, and mysticism". *Sociological Analysis* 34(2): 140–149.

Turner, V. 1986. "Dewey, dilthey, and drama: An essay in anthropological experience". In *The anthropology of experience*, eds. V. Turner and E. Bruner. Urbana/Chicago: University of Illinois Press, pp. 33–44.

Weber, M. 2005. *Protestant ethic and the spirit of capitalism*. London and New York: Routledge.

Zhouri, A., Laschefski, K., and Pereira, D. 2005. "Introdução: Desenvolvimento, sustentabilidade e conflitos ambientais". In *A insustentável leveza da política ambiental: Desenvolvimento e conflitos socioambientais*, eds. A. Zhouri, K. Laschefski and D. Pereira. Belo Horizonte: Autêntica, pp. 11–26.

Zhouri, A., and Oliveira, R. 2010. "Quando o lugar resiste ao espaço: Colonialidade, modernidade e processos de territorialização". In *Desenvolvimento e conflitos ambientais*, eds. Andréa Zhouri and Klemens Laschefski. Belo Horizonte: Editora UFMG, pp. 439–462.

Interlude

Sociality of the chickens

Figure Interlude 2.2.1 Movement of chickens

The chickens slept at ten o'clock, as usual. This was the moment a human turned off the lights of the hennery.

It was a greyish Saturday, warm temperature, in the winter.

The chicken were "companion species" (Tsing 2012) for the humans who lived in the smallholding Muriqui Assu. Observing their social relations

DOI: 10.4324/9781003378853-5

among them and with humans is a daily exercise of balance between human and non-human. Knowing their sociality is to take in full consideration the practice of permaculture as an "ecological epistemology" (Carvalho 2016) in which relations with non-human beings are constant. In this shared environment, cycles of reciprocity and flows of materials are entangled through the engagement in lifeworld.

Besides the analytic complexity of describing more-than-human life – because such an ethnographic practice requires attention to the forms and language of other beings who do not talk, challenging the notion of intentionality and the categoric imperative (Tsing 2013) – I would need enormous prudence to sit among the chickens, if I wanted to observe their sociality. The chickens were big, healthy and well-nourished. They were taken care of daily and spent a good portion of the day free. Wilson allowed me to observe them.

Samuel, a black man in a vulnerable situation, worked part-time in the smallholding. He warned me, his face serious, that I should be wary and prevent the chickens from escaping. I opened the first wooden gate, shooing them with my hands, so they would keep inside the hennery. I locked the gate latch. I came inside with a bucket in which humans usually brought food. I had already fed them earlier, as soon as we woke up, like always. I wanted them to understand there was food in the bucket, and then they would stay there, without trying to go out, because they always did.

I came into one hennery, and I stayed still for some time. I squatted on a corner and stayed there, observing.

Some chickens approached the food tray, pecking the remains of sparse grains, simply repeating their common gesture before the human presence. From humans, the chickens expected a few things: feed them, free them, lead them back to the hennery, capture the escaping poultry and take their eggs. Some chickens also looked at me, approached me, but they never let me touch them. One of them drank water. With her little eyes, she observed my small quiet movements.

To my surprise, a group of chicken managed to push the door, opening the latch. I stood up rapidly to prevent others, dozens, from escaping. Samuel was watching everything. And he did not look happy.

I stayed there observing the escaping chicken scratching the ground, outside the hennery. Some pecked the bucket, looking for bran. Others approached me. Others tried to fly outside the wall, opening their space for an inopportune hike, caused by the disturbance of an inopportune human.

Figure Interlude 2.2.2 The revolution of the chickens

References

Carvalho, I. C. M. 2016. "Ecological epistemology (EE)". *Encyclopedia of Latin American religions,* ed. H. P. Gooren. Switzerland: Springer International Publishing, pp. 1–3.

Tsing, A. 2012. "Unruly edges: Mushrooms as companion species: For Donna Haraway". *Environmental Humanities* 1(1): 141–154.

——— 2013. "More-than-human sociality; A call for critical description". In *Anthropology and Nature*, ed. por Kirsten Hastrup. New York: Routledge, pp. 27–42.

3 *Gaudya Vaishnavismo* in Serra da Bocaina (Rio de Janeiro)

Figure 3.1 Radiant fields

Devotees achieve a different perception of the environment when they reach the understanding of the world as an emanation from Krishna – the things and beings in the world have soul and energies that emanate from the deity The people who are dedicated to *Bhakti Yoga*,[1] i.e., yoga of devotion, perceive the phenomena of the world with equanimity because they understand that everything shares the same soul. The vegetarianism of the vaishnavas (devotees of Vishnu, another name of Krishna) is a simple expression of compassion

DOI: 10.4324/9781003378853-6

towards the soul in every being. Nevertheless, most vaishnavas do not dwell in ecovillages.

The Hare Krishna spiritual practice, everywhere, is experienced as a devotional service whose aim is to establish a personal relationship with Krishna. He is understood as a person whose consciousness (*Purusha*) enjoys the phenomena manifested by his own energy (*Shakti*). His energy, which is his soul, impregnates everything. The individual souls are infinitesimal whilst Krishna is infinite. Individual souls are to Krishna what the drops are to the ocean. Every action of the person in relation to the world can be sacralised because the person dedicates everything as an offering to Krishna, who is, according to Vaishnavism, the owner of everything. Thus, there is no profane action, for the action aprioristically profane becomes sacred and yet another form of offering devotional service to Krishna. Even though the material reality (*prakriti*) is conceived as illusory, due to its temporariness, vaishnavas do not reject the necessary action in the world.

Those concepts are expressed by vaishnavas, the devotees who live in *Goura Vrindavana*, an ecovillage associated with ISKCON (International Society for Krishna Consciousness). Such conceptions enable the understanding of Vaishnavism, and its style of spirituality,[2] as epistemology. My understanding of this epistemology reflects the fieldwork process, being based on the evidence that vaishnavas, in different places, share the same cultural premises. *Goura Vrindavana* and other Hare Krishna ecovillages – such as *Vrinda Bhumi*, which will be observed closely in the following chapter – share the same premises.

The ecovillage *Goura Vrindavana*, located on the top of the mountain Serra da Bocaina, is 12 miles from Paraty city centre, state of Rio de Janeiro. About 300 metres above sea level, the land in which the ecovillage was built has unbelievable 800 hectares. The dwellers do not have even 100 hectares of it used or even cultivated. An enormous portion of land became part of the forest reserve Parque Nacional da Serra da Bocaina. From the ecovillage, the archipelago and an almost green sea inlaid with stones and hills can be seen. To reach the place, it is possible to take a common bus at the bus station in the city centre. From the foothill, it is possible to walk for 1 mile, through a dirt road surrounded by rainforest, enormous rocks and bodies of water flowing down the hill.

It was a rainy afternoon when I arrived at the ecovillage. At 6 p.m., a specific funerary rite was held at the temple. To reach the temple, which was situated at the highest point of the ecovillage, one needed to walk by long stairs.

A devotee had died and even though his body was not there, the rite celebrated him. Shesha, a devotee who led the rite, said that generally the bodies of devotees were cremated, and the ashes scattered on a sacred place, like seas or forests. Nevertheless, Shesha did not know the destiny of this body because the family of the devotee had another religion and they could prevent

Figure 3.2 – Temple

the cremation. Shesha, whom I saw only once, had experience with leading the funerary rite. According to him, the rite invited the person (*jiva*, individual soul or consciousness) who abandoned the body to come to the temple, listen to the reading of a sacred text and prepare the next steps of the travel. There was chanting of the *Mahamantra* Hare Krishna and playing of *mrdanga* (ancient Indian drum struck by the fingers on its extremities, always present in vaishnava temples).

Hierarchy and power?

The *arati* (rite of adoration and offering; same as *puja*) always happened at 4.30 a.m. In the temple, there were three women and a young man, standing up and forming a semicircle before the altar. After the rite, devotees would take pads and sit on the ground. This new semicircle was formed before the guru or someone else who was reading a sacred text. That morning, Vraja-Lila – a devotee, 32 year old, biologist, coordinator of the ecovillage – was the lecturer. Vraja took a copy of the book *Srimad Bhagavatam.*[3] Sitting in lotus position, she read an excerpt of the Eighth chapter from the Tenth Canto, and then she started to explain it.

The narratives from the Eighth chapter regards the birth of Krishna, and the preparation for the name-giving ceremonies. Devaki gave birth to Krishna, but Nanda Maharaj and Yashoda adopted him, because King Kamsa, Devaki's brother, killed all nephews, so that they would not depose him from his position of power.

Naming ceremonies are very important in Vaishnavism. An important initiation rite for a new devotee is also a naming ceremony, where another name is given. The initiation is the way by which a vaishnava devotee begins the path of becoming a *brahmane.* Vraja, who led the study of the sacred text, referred to the Hindu social system (*varnasrama dharma*), according to which the social hierarchy is divided into four castes: *brahmanes*, priests; *ksatriyas*, warriors; *vaisyas*, traders; and *sudras*, servants (Dumont 1970). The social philosophy of Vaishnavism conceives that *brahmanas* are the head of the social body. Nevertheless, there is a fundamental distinction from the traditional caste system. Vaishnavas understand that the category of brahmanas includes not only people who were born in such a caste, but rather everyone who reaches such a "state of consciousness". Through the practice of *Bhakti Yoga*, anyone could become *brahmane*, regardless of the rigidity of the Hindu social system.

At the temple, Vraja was suggesting that, according to the vaishnava philosophy propagated by ISKCON, the Hindu system could be expanded to "all human society". She was suggesting the possibility of universalisation of a social philosophy embedded with the spiritual practice of *Bhakti Yoga.*

Later, when I was alone, I sat on the ground inside the temple, took the book, and read the comments of Prabhupada, the founder of ISKCON. He criticised the notion of exclusivity in the Hindu social system, suggesting the possibility of spreading the vaishnava social philosophy globally through which many people around the world could become *bhramanes*. The whole excerpt asserted that:

> The Kṛṣṇa consciousness movement is therefore very much eager to reintroduce the *varṇāśrama* system into human society so that those who are bewildered or less intelligent will be able to take guidance from qualified *brāhmaṇas. Brāhmaṇa* means Vaiṣṇava. After one becomes a *brāhmaṇa,* the next stage of development in human society is to become a Vaiṣṇava. People in general must be guided to the destination or goal of life, and therefore they must understand Viṣṇu, the Supreme Personality of Godhead [...] It doesn't matter whether one is born a *brāhmaṇa* or not. No one is born a *brāhmaṇa;* everyone is born a śūdra. But by the guidance of a *brāhmaṇa* and by *saṁskāra,* one can become *dvija,* twice-born, and then gradually become a *brāhmaṇa.* Brāhmaṇism is not a system meant to create a monopoly for a particular class of men. Everyone should be educated so as to become a *brāhmaṇa.* At least there must be an opportunity to allow everyone to attain the destination of life. Regardless of whether one is

Figure 3.3 Arati

> born in a *brāhmaṇa* family, a *kṣatriya* family or a śūdra family, one may be guided by a proper *brāhmaṇa* and be promoted to the highest platform of being a Vaiṣṇava. Thus the Kṛṣṇa consciousness movement affords an opportunity to develop the right destiny for human society.
>
> (*Srimad Bhagavatam*, Canto 10, Chapter 8, Text 6)[4]

Apparently, the commentary of Prabhupada resembles a fundamentalist discourse in which ISKCON aims to universalise the Vedic texts and the system of castes (*varnasrama*). Nevertheless, the excerpt unfolds a subversion of the Hindu notion of castes. The ideas that "everyone should be educated so as to become a *brāhmaṇa*" and "everybody is born a śūdra" involve a kind of spectrum with two extremities: the pure and the impure. Dumont (1970) articulated the Hindu hierarchy in oppositional terms: the pure is to the superior what the impure is to the inferior. In the commentary above, even though those oppositions underlie it, the position of the hierarchy is entirely inverted. If everybody is born *sudra*, impure, regardless of the family, each person would pass through the process of becoming a vaishnava (and practise *Bhakti Yoga* accordingly) to be considered pure, *brahmane*. The elevation in such hierarchy regards the elevation of consciousness.

The curious conception reminded me that when I introduced myself to Madhurya Devi the previous day, she already knew my name. Someone told her that "a devotee" would arrive and Madhurya was ready to welcome her. I had to correct the information, saying I was not a devotee. Vanessa, another volunteer in the ecovillage during those days, said: "Everybody is devotee". And Mateus, her partner, added: "Everybody can be devotee regardless of institutions". If everyone can be a devotee, then everyone can be vaishnava and reach the state of *brahmane*.

The vaishnava idea that everyone is a devotee – for all humans are considered the very emanation of Krishna, even if they are not affiliated to any institution, initiated, or do not practice *sadhana* regularly – is related to the horizontal notion of social hierarchy. It was interesting to observe how this ontological premise was manifested in the ecovillage.

During 11 days, I observed that, even though there was no evident vertical hierarchy among the dwellers, discursive clashes were common. Vraja once told me that, despite the divergent ideas, the philosophy of Krishna consciousness united the dwellers. Only devotees were allowed to live in the ecovillage. The morning I left the ecovillage, Gosta asked me if my visit was fruitful. Gosta, who lived in another Hare Krishna ecovillage, said: "We have everything here. It lacks only spirit of community. It is each to his own and God watching over everyone". The individualism was generated by the devotional practice, which absorbed each person in their continuing process of Bhakti Yoga, the Yoga of Devotion to Krishna. Yoga in Sanskrit means union. And such a union was the objective of every activity.

Each person seemed to inhabit a personal world in which every activity was seen as an opportunity to offer devotional service to Krishna. Sometimes, communication was diffuse, and each person followed their own logic. I remember that I was taking care of the garden next to the kitchen. My task was removing some carurus (*Amaranthus sp.*), a non-conventional food plant. I deposited the leaves on the compost, which was generating manure for the garden. Rupa arrived and helped me, without being told to do so. And then he helped Mahendra with his little tractor, with which he transported bamboo for his bioconstructions. Manoel, an elderly man who worked in the ecovillage for many years, asked me to pluck some tomatoes from one garden plot.

Gosta was cooking lunch and he saw me plucking the tomatoes. He looked at me with an upset stare and asked me to leave the tomatoes there. I had to replant the tomatoes in other soil.

During the meals, most people who lived in the ecovillage joined together, chatted and commented on the situations involving each other. Affinities were expressed during the meals. And the topic of conversation was often nourishment and food. Some devotees were vegan (strict vegetarian, who do not eat anything provided by animals) and very attentive to food, and to its properties and functions as part of the menu.

Figure 3.4 The living garden

The cooker blew the sacred conch whose sound reminds of the primal mantra *Om*. Lunch was ready. We sat around the table for the *prasada*. Prabhupada, the Indian erudite who brought Vaishnavism to the United States, always emphasised the importance of eating *prasada* and food offered to Krishna. The moment of the meal was delicately ritualised in a formal fashion.

Mahendra expressed his point about the textured soy protein over the mashed potatoes. Mahendra said that soy protein was not human food, but the residue of the soy oil that also serves as fodder. He did not mention the economic aspect of soy monoculture in Brazil. Madhurya, who had been a vegetarian since she was born, said she had always eaten soy and was currently in perfect health. Mahendra answered that soy was not genetically modified before. And Madhurya said loudly that Mahendra and other people imposed their visions sometimes. The altercation was uncomfortable for everyone sitting around the two shared tables. Mahendra accused Madhurya of being daring. And she replied: "I dare expose ideas different from yours – what other people don't. I dare to question your booklet". At night, during the meal, Madhurya returned to the topic of food radicalism and Mahendra, who lived in a wooden house above the stable close to the kitchen, listened and came down saying that he understood the situation. More debate followed.

The faces demonstrated discomfort. Madhurya said: "It's because we love and hate each other for a thousand years". Mahendra responded: "I do not love you". An appalling silence filled the space and the group dispersed.

It seems that the notion of the horizontality of social relations engenders a wide acceptance of individuality and personal freedom, as well as facilitates power relations. Power relations existed, but they were alleviated exactly by the respect for personal freedom. In the ecovillage, the argumentative conflicts did not explicitly aim for agreement or acceptance of different positions and ideas. There was exposure to divergent ideas without the intention of changing positions.

Drawing from her fieldwork research on the New Age, Amaral (2000) suggested that alternative communities tended to be "apolitical", not focusing on differences that could cause oppositional clashes. Nevertheless, it is impossible to assert the absence of a political life in alternative communities, even if the clashes happen on the micropolitical level. Why does a monk, who is always in contact with deities in the altar of the temple, feel comfortable to consider other people impure or analogous to animals? The issue at stake is power relations.[5]

Later, during our interview, Mahendra told me that "social life is an opportunity for self-work. Living in the ecovillage is not only meditation, peace and love". According to him, "if two devotees altercate and express words aggressively, and if they are on the correct disposition, then they will have a different vision. Why did someone say that? Krishna is trying to show me something". The dispositions can be of three kinds, or modes (*gunas*) in which the material energy unfolds: *tamas* (ignorance mode), *rajas* (passion mode) and *sattva* (goodness mode). The concept of modes is also embedded in a theory on personalities and humour. According to vaishnava philosophy, the ignorance mode (tamasic) is relative to the darkness, decline, absence of movement; the passion mode (rajasic) is the other pole, the movement, pulsation; and the goodness mode (sattvic) is relative to the balance between the other two. The "correct disposition", as an abstraction, would be related to the moment.

The experience of sharing the same place with other people seemed challenging. Different minds, immersed in their own particular interests, encountering the interests of others. Each person has necessities and the dilemma seemed to be letting others freely meet their needs. Even though such relations were not necessarily conflicting, there was indeed a clash between the individualistic action oriented to one's own needs and the altruist action oriented to the necessities of others.[6] While the first could become a factor of attrition in social relations, the later was considered positive. An example: during the meals, some people took the plates first, but let others get to the table before (*sattva*); some people let others come to the table before (*tamas*); and some always cut in line (*rajas*). These three forms of being served demonstrated their current state of humour, which became the humour of social relations happening at that specific moment. The trend was following the impulse for food or controlling the impulse for a moment, letting others be served first. The dilemma was silent.

Spirituality as search for connection with nature

Vanessa invited me to go to a waterfall after lunch. The couple was on this journey of exploring ecovillages. After *Goura Vrindavana*, Mateus and Vanessa would keep travelling through the Brazilian roads. They were always gentle. Later, I recorded my dialogue with Vanessa.

The chosen place was one of the waterfalls near the lodging, inside the ecovillage. When I arrived at this waterfall called *Bhakti Kunda* ("river of devotion"), Mateus fingered simple chords on the acoustic guitar, Vanessa blew into a flute, and both sang. "Mother Divine I want to be/A realised child/ Before your power/I surrender to be released". They gave me a little drum, which I began to strike. Mateus intoned religious chants, neoshamanic and umbandistas. It was an artistic moment that Mateus called "caboclagem", suggesting he was invoking the forces of spiritual beings of the forest (which are called "caboclos" in Umbanda and other African-Brazilian religions). Even though the context apparently was not religious, the moment was not only religious, but also spiritual for the couple. The chants were prayers and a form of expression of this search for connection with nature. In an instant, the stone of the waterfall became a temple in which the sacred erupted. It was a manifestation of the immanent sacred in nature, revealing energies, forces and flows that have always been there, often unnoticed by those who did not learn how to create such a connection.

The spiritual experience of connection with nature is rooted in a sacred perception of the environment (Chiesa 2017) where all beings, humans and more-than-humans, are entangled in the web of life. To Vanessa, spirituality means "living in integration with nature". This notion of spirituality as the search for connection with nature had different aspects.

Vanessa's voice crossed the silence: – When a person is in nature, the healing is felt. The person feels the purity of the waterfall, the purity of the ocean that purifies them. So, this energy, this purer matter, has not been in the hands of a human with a confused mind. It has been in the hands of beings of light with pure minds.

Besides the ontological statement on the reality of incorporeal beings that inhabit nature, Vanessa understood the human experience of connection with nature as a therapeutic process. This process, experienced daily in places such as the waterfall, reminds me of the idea that the sacred is intimately related to the healthy (Bateson 1991), as well as to the aesthetic, beauty and search for patterns in nature. The sacred perception of the environment is attentive to the continuity and symmetry between all beings and can include the integration of the mind as a spiritual experience. Such a perception is reached

Figure 3.5 Vanessa and Mateus

through a specific modality of attention (Chiesa and Gonçalves Brito 2022), which involves meditative exercises in restorative environments. Developing such a perception seems to enable the preservation and restoration of direct attention skills, impacting mental states positively (Kaplan 2001). The idea of "being present in the moment" – important to different spiritual trends of thought – is a possible analogy for this healthy mental state of attention to the flows, movements and continuities which unfold in the engagement with the world.

The evening came amidst high trees and densely wet air. Vanessa was receiving me on the wooden house where she and Mateus had been staying. We could hear cicadas singing happily in the dark. We could also hear the sound of water that uninterruptedly circled the structure of stone, which was a rustic washing mechanism.

We sat on the ground, and she told me she was 20 years old and she was born in a very small city of the state of Santa Catarina, whose name was Xaxim. She told me she was dedicating her time to self-knowledge. Mateus and Vanessa would be guests for two months. *Goura* was the first ecovillage they were visiting during their backpacking journey through the roads of the Brazilian coast.

I asked her if they planned their trip and how they had decided to travel. Vanessa replied:

– It was a surrender. But I have spiritual goals too. I did not plan because I really decided without any planning. I have never thought it would be so soon, young, without money, without formation. Simply going to the world. And learning how to make my art and my work in life. It was really a huge surrender because who will supply me but the Great Spirit, on this way? I spent some time into the wild before, meditating and learning how to survive in the forest, taking some time to myself. I have learnt I can be very well anywhere, learning to survive. And then I went to the world. I want to evolve, expand my consciousness, illuminate myself. I want to find the cosmic kinship, the family I can share, with which I can illuminate myself and illuminate the others. Give hands to form a great chain, united. I felt I needed to make this energy and consciousness circulate through Brazil. Seize this opportunity of having a fertile mind now. Very open to new knowledge and various cultures.

The spiritual goals mentioned by Vanessa pointed to a variegated learning process, through a travel that resembles the countercultural ideal of beatniks and hippies. Both (counter)cultural movements happened as a response to the hegemonic lifestyle that the keen youth could never simply accept, in their youth times. This *zeitgeist* still pulsates by the roads of Euro-American countries. Vanessa and Mateus were living examples of people who move through the cities around the world, taking and bringing with them concepts and techniques – artistic, spiritual and scientific – like birds who disseminate seeds.

The words of Vanessa also revealed a diffuse practice of New Age circles, which is the transit of autonomous travellers in different collectives and communities, and the consequent exchange of knowledge. Some travellers, such as spiritual nomads, can create an individual synthesis of different concepts and practices. Notwithstanding the cultural differences, these people consider each human as part of a universal family and the reciprocity between them crosses global webs of exchange. Such exchange is not necessarily based on a monetary logic, but rather on shared ethics and concepts that transcend the frontiers of national states. On the other hand, technology is not always fundamental to exchanging, whilst bodily travel is still a core tenet in the construction of such narratives. These people are "pollinators" of a global consciousness and a cosmic spirituality, ritualising symbols and concepts from diverse and divergent cultures, be it Celtic, Indigenous, Sufi or esoteric (De La Torre 2011).

Observing how diffuse and permanent was the influence of the countercultural ideal, I asked Vanessa to tell me about some experience with nature, which was particularly important to her.

With a serene voice, almost whispering, Vanessa replied: – I had not had a specific experience that brought me the biggest cure or expansion. But rather a constant search for connection. Today I am fasting. I sat alone, talking with the trees. I talked with the earth. And she showed me many things. I saw them clearly. Many things we already know and sometimes forget. We are soul, the Great Spirit. God is soul, a Great Spirit in everything. And everything has parts of its soul.

In the context of Vanessa's idea, the spiritual experience of connection with nature expressed the continuity and symmetry between nature and herself. The spiritual experience also revealed a kind of pantheism according to which God is conceived as a monad or spiritual particle infusing everything. At the same time, there is also a monism in which the particular souls of beings form a magnificent entity, which is the very Great Spirit. Curiously, Vanessa understood the multiplicity and unity of the Great Spirit. Such an idea simultaneously links monism and pantheism in such a way that they do not exclude each other, even if they are philosophically incongruous. This simultaneous juxtaposition of conceptions pointed to the personal freedom of choosing spiritual experiences, but also some ideas of neoshamanism.[7]

Vanessa said her experiences of connection with nature also included ingestion of entheogens such as ayahuasca, which is a millennial beverage used by Indigenous people of the lowlands in South America. The beverage is the result of the decoction of the vine *Banisteriopsis caapi* and leaves of *Psychotria viridis*. To Vanessa, the contact of ayahuasca was a healing process.

She said: – Ayahuasca itself teaches what you need to do, the next steps of your journey. Like a mother who is sometimes rigid, but who tries to teach. If you are open to learn, you learn everything that comes from nature. The healing… Each plant has its own spiritual mission. As well as human beings. And it seems I already knew that all this backpack travel would happen exactly as an opportunity to heal some things and connect with nature. Ayahuasca is a teacher who teaches how to connect with the forest beings, respecting nature. She purifies the vision, so that you can look beyond what the eyes can see. See the soul. See the soul in others and perceive you do not need synthetic or industrial things. Only nature itself is sufficient. It is complete in itself. Humans separated from nature when they began to think they were superior, modifying nature and producing things. But we need to return to be whole and wild. Not exactly wild, but tribal.

Vanessa expressed a certain utopian ideal of a numinous past when humans lived intimately with nature. Or when the humans were purer and still not affected by modernity that made them believe in their separation from nature. The healing mentioned by Vanessa was her personal overcoming of the ontological disjunction between nature and herself, as a human. Vanessa also introduced a common idea in circles of ayahuasca.[8] The idea that the sacred beverage is itself a source of knowledge (Oliveira and Boin 2017). For the epistemology of ayahuasca circles, the ingestion of the entheogen – etymologically, a plant that contains God – enables a process of apprenticeship only experienced by means of contact with the beverage. On the other hand, the attribution of intentionality to ayahuasca as a being that teaches humans challenges the Cartesian assumption, which divided the conscious human being from nature as an inert object. Such a notion also dissolves the ontological discontinuity between humans and non-humans. Thus, according to the premise of the epistemology of ayahuasca circles – human and non-human "should be understood as part of the same socio-cosmic field" (Oliveira and Boin 2017: 226).

On her part, Vanessa was trying to learn how to survive in the forest without the necessity of depending on the capitalist system in the cities, which she called "Babylon". Vanessa had told me before about her experience living close to trees, water, in the middle of the rainforest. She began to seek for another way of life after camping on a place called Lagoinha do Leste, in Florianopólis, a southern city of Brazil. When she came back to the city, she decided to live by the lagoon, near the ocean. Vanessa got rid of unnecessary things and left.

Responding to my curiosity and interest, Vanessa told me the details: – It was a place through which many different people passed. And I was very connected to myself and nature. Every day I drank pure water from the waterfall that drained into the lagoon. I think this made me more sensitive, sensible.[9] This contact with water. I stayed there camping for eight months, meditating a lot. I feel touched when I remember the moments I lived there. I want to learn about ecology, agroecology, permaculture, and then I can live totally integrated with nature, without the need for external things. Just me and nature because it is very pure, extremely beautiful.

– When you say you want to live a simple life connected with nature, without depending on the capitalist system, it seems you are distancing yourself from the world. But you are not actually distancing, you are thinking on the world, I remarked.

Vanessa agreed: – Yes. I believe we do not have to destroy or struggle anymore. We need to accept, recycle and transform. Transform this world. This is also a form of transcending. Transcending the material. Transcending the desire for changing people. We have to do our part and build a new way of living, a new way of being here. So, when people awake, they will find this new world and see it. People stay where they want, but when they search for it, they will find it, as well as I will find it too. This consciousness, this knowledge of how to build the community. My political vision is communitarian. I do not think that people need a leader. I think that we need to live on community and learn how to live as community. Maybe various alternative communities, integrated among themselves, with everybody learning the laws of the universe, the universal laws.

The possibility of a social transformation that happens from inside out has always seemed intriguing to me, all the numberless times I heard versions of this concept from many different people. But Vanessa also had her own concept about this process of change.

– Within, changing and healing oneself. It is the only way to heal the world. First, transforming within all this illusion, transmuting all this negative thought into love, joy, peace. Then, you will share it. All energies you emanate will return. The change begins within each person, generating a collective consciousness. Thus, we can build the new earth for the future generation to come.

– How could the transformation of a person change the world? I inquired, touching the notion of spirituality as transformation of the self and world.

With open, wide, vivid eyes, Vanessa replied: – Without any imposition. How do you share the change without imposing? I think that people can be open to see this light. And to observe too. They can observe you and feel what you have to transmit. You do not need to impose, people will ask you, they will sit by your side, be with you and listen. Sooner or later, someone will feel and search for this. And you will keep your transformation, your evolutionary process. People will remember you and see you somehow. They will follow your process and see the realness of your words. They will know you applied your words to yourself. You transformed yourself. This is how the revolution comes.

To understand this notion of the inner transformation that reaches the world, it is necessary to understand the entanglement of person and world. Only a monist concept of the relation between individual and society can resolve Vanessa's paradox. First, it is necessary to understand that beings – including human and non-human beings – are always immersed in this relational matrix of existence (Strathern 2005). In social life, there is never an individual isolated from the world because the life-worlds which human and non-humans inhabit are always social worlds, as I learned from the studies of Schutz (1967) and Schutz and Luckman (1973). Even the monk who decided to live in a cave has probably renounced the world, which exists independently, to dive into his inner reality. But if he refused the world, it does not mean that he left life in the social world. The distancing from the world is just the affirmation of the undeniable reality of the world. The monastic life of the monk or the meditative experience of people like Vanessa are not simply personal. Moreover, if the person is already and always immersed and entangled with the world, the inner experience will inexorably reveal the world within. The problem of such a relational perspective – notwithstanding the relevance of its monist concept of the social world – is that the notion of relation often presupposes beings and things as separate prior to the establishment of their connection. Ingold (2008) noticed that the mutuality between parts can only be explained by the inversion of beings and things upon themselves, as if they were self-contained. Therefore, Ingold (2008) suggested moving beyond this idea of connectivity between things and environment, in order to think in terms of lines which entangle beings and things in the fluid space, paths of flow "like the riverbed or the veins and capillaries of the body" (11).

Each being is, thus, not only a line, but a thread of lines, a gathering of threads of life (Ingold 2008: 12). Social life emerges, contingent and indetermined, through the flowing entanglement of threads of life, which form a meshwork in constant becoming (Ingold 2015). The lifelines that entangle beings, things and environments are always responding to each other (Ingold 2017), in such a way that people's actions emerge of their mutual engagement in the world. The "relations" of people – their lifelines – are correspondences that affect and constitute their existences in permanent becoming. The attentive self who perceives such correspondences is transformed, and then, their movements also can move the threads entangling the movement to other lifelines.

The entanglement of the lifelines of beings and things in the world does not happen through mere interaction between them. According to Ingold (2015),

> "Between" articulates a divided world that is already carved at the joints. It is a bridge, a hinge, a connection, an attraction of opposites, a link in a chain, a double-headed arrow that points at once to this and that. "In-between", by contrast, is a movement of generation and dissolution in a world of becoming where things are not yet given – such that they might then be joined up – but on the way to being given.
>
> (147)

Following this insight, I would suggest that the transformation of people and the social world in which they live happens without a prediction, a final destiny. It emerges from the flows that move the entanglements of the lifelines of beings.

Even so, the notion of the spiritual experience as the search for connections between parts which were separated – even if they were not always

Figure 3.6 A cow looked at me

ontologically divided – only makes sense in the context where humans were really separated from nature. Indeed, such a separation that requires posterior connection is an ontological deviation of the modern Euro-American thought. An epistemological error, like Bateson (1972) suggested. Maybe Vanessa was mentioning exactly this problem, when she talked about illusion.

Besides taking care of the garden, once I was called to help with the pasture that would receive the family of bovines who inhabited the ecovillage. Mateus, an enthusiastic 19-year-old boy, handled the machete. I took the scythe. Mateus and I had a conversation whilst we were working. And he told me of an ecovillage in the state of Paraná where people lived for many years without electricity, eating organic plants, with their feet literally on the ground, intimately connected with the earth. These people, Mateus said, criticised the "system". We questioned if the world "system" is adequate to grasp life, including living organisms, which are often described as living systems, or organisms with their own internal systems, such as the circulatory system, respiratory system and so forth. Mateus exposed his ideas about the world system, and he thought the world was negative, especially from the social and political perspective.

His ideas reminded me of the reflections I had had for many years, and I wanted to share with him. I asked him to listen attentively because my intention was not to convince him.

I told him that, according to Maturana and Varela (2003), organisms have been understood as living units or systems which pass through transformations and try to keep themselves alive. I explained the concept of homeostasis according to Lovelock and Margulis (1974) and told him that the notion of system was used by thinkers like Bateson (1972), who was a revolutionary by his time. Even earth, I said, could be understood as a unit, a system or living organism which is constantly maintaining the balance of its constitutive elements, or ecosystems.

Mateus, thinking on the analogy, said that if Earth is an organism, then some unbalanced cells have been hampering the functioning. Stretching the notion of system and borrowing his metaphor, I said that if the social system can be imagined as a living organism, one could say that some sick cells of the social body are the people who want to transform the system, in order to break its homeostatic process. Nevertheless, one could say that the social body is sick and the inept cells, which cannot adapt to the homeostatic process, are like the white blood cells, acting to heal the rotten infections through the action that generates other form of life.

We stretched this naturalistic metaphor even more. I suggested that the social system would "function" in a certain way to benefit (and maintain the privileges of) some small portions of the global population. Mateus and I discussed the reasonable critique of the theory according to which the system is functional. The functionality of the system, we thought together, was evidently a justification for the maintenance of elite power, as if it was the natural course of life.

Mateus and I understood that the concept of system is not positive or negative per se. Its meaning becomes different, depending on the people or collectives that use it for their own purposes.

We were cutting trees, pruning plants and preparing the pasture to receive the cattle who were coming to be fed there. Reaping the plants in the highs of a field in Serra da Bocaina. Apparently, the scythe had two meanings. It could be the symbol of death. But, if it was observed from a different angle, in the context of agriculture, it means transformation.

Common days in the ecovillage

Not everyone who stayed in the hotel inside the ecovillage frequented the rites; nor was every volunteer interested in the philosophy. I tried to attend at least one out of two daily rites. Either the *Mangala Arati*, at 4:30 a.m. or the *Goura Arati*, at 6 p.m. I always seized the opportunity to read chapters of the sacred books available inside the temple.

On a certain morning, I was asked to reap the excessive plants in the piece of soil behind the temple. People wanted to plant another specific species. I also pruned coconut leaves. Then, I bathed in the stream and, when I was walking to the refectory, I met the couple Rupa and Tulasi, who came to talk with me, and already knew my name. Rupa was born in Portugal; he was approximately 40 years old. He said that Madhurya had told him I was an anthropologist who was there as a researcher. I succinctly explained some ideas I was exploring in my dissertation: (1) – the notion of spirituality as an experience of transformation which is not restricted to inner life; (2) – because when one reaches such a spiritual experience in connection with nature one reaches a different perception of the more-than-human beings; and (3) such a perceptual shift can resonate in the social world to the extent people act differently due to their spirituality.

That night, the rite was very ebullient. First, the pujarini (a priestess devotee who is responsible for leading the rite) offered fire to the deities at the altar. The fire was brought to each devotee, who raised their hands towards the little flame. Then, water was offered and brought by a devotee, who spilled some drops over the heads. At that moment, people were happily dancing, with arms above their heads. A circled was formed. Women went to the right and men to the left. A flower was offered by the pujarini and Gosta brought it to the devotees, so everybody would feel its fragrance. The women clapped their hands, whilst the men played instruments, like bells and mirdangas, which are small Indian drums. After dancing and chanting, a sacred text was read by a devotee. Rupa chose the Eighth chapter of the *Bhagavad Gita*, according to which absolute truth is a person, Krishna.

After that, I asked some questions about the vaishnava philosophy. Rupa remarked an episode where Sri Caitanya Mahaprabhu, one incarnation of Krishna, converted *sannyasins mayavadis* (monks who followed nihilist philosophies or *Advaita Vedanta,* the non-dualist Vedanta) to the personalism of Vaishnavism.

Rupa was interested in understanding how I was articulating ecology and spirituality. He had lived in the ecovillage for many years, and he noticed that many vaishnavas do not have a close relation with nature. Some devotees live in concrete houses of cities, and some do not even know what bioconstruction is.

I explained that the notion of nature, which is the epistemological foundation of certain anthropological theories, pertains to an ontological substrate shared by living beings. Human beings separate themselves from the biological level through their capacity for generating culture. According to certain theories, that ability led the human being to this separation from nature. Nature is then considered by humans as everything that supposedly lacks a form of intentionality or consciousness enabling them to think about themselves like humans do. I emphasised the nuances regarding the anthropological notion of culture. I also told him that part of my work was trying to break the ontological division between nature and culture, human and non-human, returning to an ecological epistemology that considers the continuities between all life forms (Steil and Carvalho 2014). The search for this symmetrical observation – not only of the relations, but also of this mutual engagement of every being in a shared world – is a form of ecology that becomes entirely congruent with the notion of spirituality, wherever it points to the search for reconnection of people with a totality. Many people with whom I dialogued mentioned this sense of totality, be it in planetary, systemic, energetic, or cosmic terms. They also told me of cosmic consciousness, universal soul and so forth. Even Mahendra had told me that the mutual influence between people and the environment was energetic in nature. All beings would share the same energy and consciousness which exist in different forms. Every being would be a cell in the body of God, responding to each other's movements. To Mahendra, each being is infused with consciousness.

Rupa seemed very astonished by these reflections. The next day, we removed carurus and clovers that stifled the arugulas and lettuces in the garden. The following morning, I was astonished when I perceived the movement of the carurus, whose stems shifted to a different position. They were removed but their little roots were not the way they had been left. They were sticking to earth, urging to penetrate it again and realise homeostasis. Which force or consciousness propelled those little stems and roots to the maintenance of their lives? The situation reminded me that the anthropologist and biologist Bateson (1979, 1991) has written, like Lamarck, on this mind that regulates the evolutionary processes of every being, working by patterns that connect all beings in this sacred unity of the biosphere. The perception of the movements of carurus also reminded me of Bergson (1922) and the notion of the creative energy as élan vital (vital impulse) that propels evolution. Bergson called this force of expansion and reproduction of life, spirit. Ecology, in a sense, is to spirituality what an ecological spirituality is to an ecological epistemology.

During my dialogue with Mahendra, I understood that my perceiving of the movements of caruru stems – withdrawn and left above the soil –was only possible by this attentivity, which happened as a meditative experience.

Education of attention (Ingold 2015) in the context of intimate contact with earth, nature and other lifelines implies not only a perception of the continuities between humans and non-humans, but also a bodily sense of symmetry among the different beings. Such a specific attention is a sensory experience that Chiesa (2017) has called "sacred perception of the environment". This perception is not merely a skill that can be developed by different means. It is not an end one could achieve through personal dedication to certain techniques. Attentivity is properly an experience.

Some episodes narrated by Mahendra can elucidate how such an attentivity happens.

The first, a hike in the forest, where Mahendra led what he called "contemplative meditation without judgments". The idea was avoiding the trend to label everything humans perceive in the environment. Avoiding a way of knowing rooted in the categoric imperative that encapsulates and dissects the material reality. Instead of categorising birds and trees, like a researcher would like to do, the proposition was overcoming speculation of the families and species of birds and trees, questioning what is to be a being tree, what is to be a bird.

The exploratory hike of Mahendra was nothing but an exercise on phenomenology of nature, suggested by Goethe (Chiesa and Gonçalves Brito 2022). A modality of thought that surges from the accurate perception of things and beings in the world. The observation of the beauty of a tree is possible exactly by the open imagination of the observer. In this process of perception, the environment reveals itself with precision because the prejudices and categories do not blur the perceiving mind. The surface of the name of a tree becomes unimportant before the moving leaves, the sap overflows around the trunk, the insects walking up and down on the branches, the circuits of roots communicating with other trees. Mahendra was experiencing the much-dreamed *epoché* of the phenomenological philosophers.

The second episode was the situation where Mahendra was planting fruits in the farm he inherited from his father. It happened years before, on an out-of-the-map region called Moinho Mirim, in the state of Minas Gerais. As a young adult, Mahendra had comprehended that planting with consciousness would bring him integration with the very movement of life. While he was planting, he remembered the conception of the fruits as a gift from the tree. The evening came and a fine rain moistened his face. He did not care because the time he felt invited to plant was exactly the rainy weather. Mahendra told me:

– I was busy planting between the coffee tree, meditating about the animals that would come to feed. Many fruits, especially bananas. Almost every animal likes bananas. I was planting lovingly in connection with the earth. When I lifted my head, there was a bush dog, near a stone, in front of me, observing me. I had never seen one there because they were rare. And the bush dog stayed there, looking at me. And I was looking back, delighted, seeing the beauty of a totally wild animal. The bush dog looked at me with consciousness, eye to eye. I tried to share with him what I was thinking. I was planting that food for him, and other animals. We stayed there for some time observing each other. I thanked him for the contact and I kept planting other remaining seedlings. And the bush dog followed me, closer. I could have reached out to touch him, but I did not want to invade his space. The bush dog was completely wild, but also friendly. When humans are in this contemplative state, we can make these connections. But our minds are turbulent, full of many concepts.

In contact with earth, this was the most memorable experience to Mahendra. Even though Mahendra had worked with earth since he was a kid, helping his father, he discovered vegetarianism and bioconstruction later in life. He then started to cultivate a sacred perception of the environment.

Once, with a child to raise, he tried to explore the field of selling and repairing motorcycles, when this activity had a growing market in his region. Mahendra participated in a course of mechanics in São Paulo, but he noticed that the coldness of the pieces was not even slightly similar to the natural elements he worked with when he planted or built recycled wood furniture.

When he returned to the farm, Mahendra started to cultivate a big garden. There were more than 25 species of vegetables. Neighbours came to ask for vegetables because the produce was abundant. There were lots of tomatoes and Mahendra started to dehydrate tomatoes. He started to produce banana flour, which was the material for homemade pasta. Later, he offered his organic produce to feed the children in the schools and started a partnership with the prefecture. His contribution to the city demonstrates how the spiritual experience encompasses a transformation of perception that leads one to act differently in the world.

Nevertheless, a point came when he had economic aspirations. If he harvested kilograms and kilograms of beetroot using only composts and natural fertilisers before, now his preoccupation with deadlines and quantity was hindering the flow of the garden. Before, he planted the seedling with a sense of gratitude, talking to her and thinking, "One day I will harvest you". Now, he was anxiously controlling the growth of the seedling. Earth stagnated, even though he was using the same methods.

To Mahendra, the commercial vision interrupted the movement of life and the abundance of the garden. Plague and disease came. The expectation of the usufruct of life for his own benefit disrespected the life cycle of the plants. He understood then the damages done by the big monocultures, which seek only profit. When the human being treats earth as a non-living thing, the possibility of biodiversity is excluded, and it enables extraction without considering the vital flows. This kind of refusal of the diverse, the small, the local and the free opens the ways through which enormous companies, such as Monsanto, establish monopolies of seeds and species for monetary speculation and profit above any ethical responsibility or sense of gratitude. The expectation of the profit enables for what Vandana Shiva called "food fascism".[10]

According to the vaishnava philosophy, Earth is a great mother, *Bhumi*, a spirit and a living being at the same time. Mahendra told me a history where an emperor in India met the spirit of the earth amid a situation of food scarcity. The spirit of the earth said that scarcity was being caused by the lack of purpose of human life. The lack of consciousness about the deity, greed and unbalance made humans perverse. The cyclical conception of time suggested that Bhumi could restrict food again, if humans became destructive again.

When he realised this, the emperor began to spread a sense of love for the soil. Then, the earth once again began to provide abundant food, and even the cows gushed milk. This allegory expressed the conception of a natural immersion of human consciousness in the whole.

When I met Mahendra, he and Matura, Vrajas's partner, were dedicating themselves to the task of setting up permaculture and agroforestry, in order to produce healthier food for the people who live in the ecovillage. Such an effort had been intermittent, partially due to the continuous flow of people who came and went. Matura would like to effectively establish the principles of permaculture effectively, applying them, and closing cycles.

An uncommon lady

Madhurya and I managed to find time to dialogue, on a warm evening. She, who was 63 years old, told me that G*oura Vrindavana* had this hippie humour. Sometimes cycles of production happened, which helped to maintain the ecovillage. For some time, there was production of dry bananas and the export of bamboo to the government of Argentina. Madhurya also told me that the lands of the ecovillage were gifted by a devotee called Setukara, 30 years prior, many years before the incorporation of the lands by the National Park Serra Bocaina.

Later, Madhurya narrated her life until then. She was born in Porto Alegre and always had contact with gypsy cards. When she was a child, there was this group of gypsies who passed through Porto Alegre periodically. A gypsy wanted to take Madhurya to travel with the group, but her mother prevented it. This gypsy woman initiated Madhurya into the art of reading cards. Madhurya left High School and attended nursing school. She worked at the Hospital de Clínicas, taking care of children in terminal stages of meningitis. She also sold artefacts her mother made. And she was married for ten years to a computer programmer. She said, laughing, that she proposed to him, and she also asked for divorce. She was an activist in Brasília during the period of redemocratisation of the country, marching for Diretas Já (Direct Elections now). She had always been vegetarian, like her parents, and frequented the restaurant Govinda in Porto Alegre, where she met Lila, who was a devotee known among the "hare krishnas" of the city. Lila shared the first vaishnava lessons with her. In 2014, Madhurya finally became a vaishnava at ISKCON in a ceremony that happened inside the restaurant.

In 2016, her father died and Madhurya moved to the ecovillage. She even left the ecovillage at a certain point, working in a hostel in the city of Paraty. She also helped the priest to prepare dead bodies for their final ceremonies of burial. She decided to return to the ecovillage because she noticed that, in her own worlds, the world did not meet her wishes. When I met her, she was

Figure 3.7 The kind Madhurya Devi

responsible for the ecovillage's gift shop, where people bought T-shirts, pants and other orientalised goods. She also offered services of virtual tarot readings. She saw herself as a good mother who took care of everybody. She said there was another woman who embodied that archetype before her.

Madhurya expressed the conception of the possibility of being in the world without "being of the world". When one has the Krishna consciousness, she said, one lives in the world without giving in to all its negativity. Madhurya was completely aware that many devotees did not live in vaishnava ecovillages. *Goura* was her family. And she could not tell how many people were part of it because the flow of people was permanent. The smile of Madhurya was always an excellent reception to everybody. Madhurya offered me a tarot reading, whilst we were talking to each other. The details of what she told me were obscured by the time was the moon who shone behind the clouds over our heads. Maybe I was sure of my new name that evening.

Notes

1 The words in Sanskrit will be presented in italic. I used a simplified form of the Sanskrit words, but I left the original graphical marks when a Vedic text was quoted. The names of deities were also typed in italic.

2 The notion of style of spirituality (Carvalho 1992) regards the proposition of a methodological assumption according to which the mystic in Brazil can be understood as an identifiable tradition. Camurça (2016) complemented the proposition with a structuralist frame. Vaishnavism in America, known as the Hare Krishna Movement, would be defined as an "Eastern meditative style".

3 *Srimad Bhagavatam* is a set of philosophical and literary sacred books supposedly written by the erudite sage Vyasadeva (III BCE?). The narratives of the book concern the *lilas* of Krishna. *Lilas* are the so-called "plays" in which Krishna becomes a bodily being and lives experiences in contact with *prakriti.* This Vedic text is also religious, and it is the core tenet of the philosophical system *Gaudya Vaishnava.* It regards various topics: from the nature of the "I" to the origin of the universe. The Tenth Canto, mentioned in the scene described above, narrates the initial part of Krishna biography.

4 This specific text and complete book can be read at https://prabhupadabooks.com/sb. Viewed December 1, 2022.

5 On the other hand, I think that the very constitution of an ecovillage – contained by the world and different from it, at the same time – can be understood as a political position. Living in an ecovillage is also a political option.

6 I draw inspiration from Weber (1978), in order to articulate the notion of the "general" social action with the kinds of social action according to the vaishnava principle of *gunas.* I refer to the "general" social action because the Weberian typology – rational action in relation to goals, social action in relation to values, affective social action and traditional social action – cannot be taken as "pure types", independent of one another.

7 The neoshamanic context in Americas is frequently associated with circuits of consumerism and cultural appropriation of the techniques, artefacts, expressions and products derived from Indigenous communities. The ingestion of psychotropic substances has always played a crucial role in the countercultural experience, dating back to the Human Potential Movement, whose epicentre was the Esalen Institute, in the sixties. An important goal of the Movement, recovered by neoshamanism, was to use various techniques to "reach the expansion of consciousness states, liberate the imagination and encourage the individual transformation and development of a profound spirituality" (Martinez 2018: 66).

8 Ayahuasca has become a powerful psychoactive substance that entangles a wide set of phenomena, such as cultural appropriation, transnationalisation of religious rites, global juridical and biomedical debate on its therapeutic characteristics, as well as the role of ethnography in the validation process of religious knowledge related to the beverage in various cultural contexts (Labate, McRae and Goulart 2010). The wide diffusion of ayahuasca is due, mainly, to the institutionalisation of its use by Brazilian religions such as Santo Daime, União do Vegetal and Barquinha.

9 This is a common notion according to which the experience of the ingestion of ayahuasca enables the cultivation of a sensitivity to nature, based on the ability to feel integrated with nature. Such integration is also a common discourse in deep ecology (Baker 2015). This idea points to another conception of spirituality, regarding the shift of one's perception. This conception of an ecological spirituality experienced as the development of a specific attention resonates with the concept of sacred perception of the environment.

10 See the following interview with the ecofeminist and ecologist Vandana Shiva in https://news.mikecallicrate.com/nobull-farming-free-vandana-shiva-on-food-fascism/. Available on September 14, 2021.

References

Amaral, L. 2000. Carnaval da Alma: Comunidade, essência e sincretismo na Nova Era. Rio de Janeiro: Vozes.

Baker, J. 2015. A thread in the vine: The deep ecology of contemporary ayahuasca discourse. Lismore, NSW: Southern Cross University.

Bateson, Gregory. 1972. Steps to an ecology of mind. New York: Ballantine Books.

——— 1979. Mind e nature: A necessary unit. New York: EP Dutton.

——— 1991. A sacred unity: Further steps to an ecology of mind. New York: Cornelia & Michael Bessie Book.

Bergson, H. 1922. L'énergie spirituelle. Paris: Félix Alcan.

Camurça, M. 2016. "Estilos de espiritualidade como critério para tipologias e interpretações do campo religiosos na contemporaneidade". Ciências Sociais e Religião 18(24): 18–32.

Carvalho, J. J. 1992. "Encontros de velhas e novas tradições: Esboço de uma teoria dos estilos de espiritualidade". Série Antropologia 131: 1–21.

Chiesa, G. R. 2017. "À procura da vida: Pensando com Gregory Bateson e Tim Ingold a respeito de uma percepção sagrada do ambiente". Revista de Antropologia 60(2): 410–435.

——— and Gonçalves Brito, L. 2022. "Learning to walk with turtles: Steps towards a sacred perception of the environment". Environmental Values 31(2): 177–192.

De La Torre, Renée. 2011. "Les rendez-vous manqués de l'anthropologie et du chamanisme". Archives de Sciences Sociales des Religions 153: 145–158.

Dumont, L. 1970. Homo hierarchicus: The caste system and its implications. Chicago: University of Chicago Press.

Ingold, T. 2008. "Bindings against boundaries: Entanglements of life in an open world". Environment and Planning A 40(8): 1796–1810.

——— 2015. The life of lines. New York: Routledge.

——— 2017. "On human correspondence". Journal of the Royal Anthropological Institute 23(1): 9–27.

Kaplan, S. 2001. "Meditation, restoration, and management of mental fatigue". Environment & Behavior 33(4): 480–506.

Labate, B. C., Macrae, E., and Goulart, S. L. 2010. "Brazilian ayahuasca religions in perspective". In Ayahuasca, ritual and religion in Brazil, eds. B. C. Labate and E. MacRae. New York: Routledge, pp. 17–36.

Lovelock, J. E., and Margulis, L. 1974. "Atmospheric homeostasis by and for the biosphere: The Gaia hypothesis". Tellus 26(1–2): 2–10.

Martínez, F. 2018. "Nueva era, neochamanismo y utopia psicodélica". In entre trópicos: Diálogos de estudios Nueva Era entre México y Brasil, eds. C. A. Steil, R. De La Torre and R. Toniol. México: Ciesas, pp. 61–83.

Maturana, H., and Varela, F. 2003. De máquinas y seres vivos: autopoiesis: la organización de lo vivo. Buenos Aires: Lumen.

Oliveira, A., and Boin, F. 2017. "O significado do daime como fonte de conhecimento". Debates do NER 32: 203–231.

Schutz, A. 1967. The social phenomenology of the social world. Evanston: Northwestern University Press.

——— and Luckmann, T. 1973. The structures of the life-world. Evanston: Northwestern University Press.

Steil, C., and Carvalho, I. 2014. "Epistemologias ecológicas: Delimitando um conceito". Mana 20: 163–183.

Strathern, Marilyn. 2005. "The concept of society is obsolete: For the motion". In Key debates in anthropology, ed. por Tim Ingold. London: Routledge, pp. 50–55.

Weber, M. 1978. Economy and society: An outline of interpretive sociology. London: University of California Press.

4 *Gaudya Vaishnavismo* in Serra da Mantiqueira (Baependi-MG)

Figure 4.1 Arriving at Piracicaba

Near the city of Caxambu, in the state of Minas Gerais, there is a small city where Nhá Chica, recently beatified by the Vatican, was born. Baependi, and its main church encrusted with gold and rubies where the imperial family celebrated masses, was not the destination (Figure 4.1). If you used public transportation, in front of a bar you would wait for the single daily bus that leads to the hamlet called Piracicaba. If someone noticed you were a foreigner, your image would be easily associated to what, in the city, they call

DOI: 10.4324/9781003378853-7

"hippies". The entire city walked before the bar and people knew that, usually, a young person would take the bus at that place, at that time. Deductively, the citizens would know if you were going to an alternative community. Many people did not see it with good eyes and then you would also feel hostility. To reach the ecovillage *Vrinda Bhumi* (Sanskrit word meaning Sacred Land), it was necessary to arrive at Piracicaba and sneak for three miles through a dirt road surrounded by short trunks and apparently dry bushes, characterising the vegetation of Cerrrado (a kind of Savana), even though the predominant biome was the remaining Atlantic Forest.

Eight young monks (*brahmacharis*) walked by a trail of the ecovillage. They wore a tuft of hair (*sikha*) on the posterior part of their shaved heads. Their foreheads were marked with a sign in the form of V, the *vaishnava tilaka* symbolising they served Vishnu. They wore the saffron clothes (*dhōti*) reserved for Indian men who followed the renunciatory life. They moved along the humid trails on a warm day of winter. The monks followed the rules of austerity related to their initiation in *Gaudya Vaishnavism*, including abstention from sexual activities for a certain time, unless they became *sannyasin* (more advance stage of monastic life) and dedicated themselves completely to spiritual practices. The monks woke up at 4 a.m., took a cold bath and uttered the *Maha-Mantra* – "Great Mantra" – holding their wooden *japamalas* inside a small bag, whose colour was saffron like their apparel. The *japamala* is a string of prayer beads used for diuturnal prayers. It allowed for concentration of the touch and for leading the physical senses to meditative act.

Most of the 30 devotees who dwelt in the ecovillage were initiated in *Bhakti Yoga* (Yoga of Devotion) through the succession of teachers and disciples (*guru parampara*) of the philosophical and religious system *Gaudya Vaishnavismo*. They promised to vocalise and mentalise the *Maha-Mantra* dozens of times a day. Usually, they could be seen with their *japamalas*, uttering the mantra whilst moving from the lodge to the kitchen. In addition, each devotee, passing through the first initiation, received a new name. The rules were (a) do not eat meat; (b) do not intoxicate the body; (c) do not practise games of chance; and (d) do not practise illicit sex (out of a marriage or stable relationship).

The monks also followed those rules and regulations. Nevertheless, in the daily life of *Vrinda Bhumi*, they were evidently different from other people. Not only in their clothes, but also in their daily tasks. For example, it was the monk's task to prepare the meals.

The kitchen was considered as sacred as the temple. In the temple, anyone could come in anytime, if they were barefoot. The kitchen was only accessible to *brahmacharyas* and *brahmacharines*, and invited people. To enter the kitchen, the volunteers needed to wash their mouths and hands.

There were two basic kinds of meal: *prasada* and *maha-prasada* (*maha* – great; *prasada* – sacred food). *Prasada* was cooked by devotees and monks,

Figure 4.2 Temple and kitchen

but *maha-prasada* was cooked only by *pujaris* and *pujarinis* (male and female devotees who were responsible for taking care of the temple). The first was food for volunteers and all the dwellers of the ecovillage, but the second was the meals for the deities.

The process of transformation of *bhoga* (common food) into *prasada* (food offered to Krishna) was always enacted by a *pujari*, who took a little bit of food out from the pan and left it in the altar of the temple.

Before serving *dahl* as *prasada* for the evening, the *pujari* lightly rang the bell of the temple door. He came in, closed the veil that separated the altar and touched another bell to awaken the deity. Deities were statues of various sizes that, according to the rite, materialised the different aspects of the Supreme Personality of Godhead (Krishna). *Pujaris* and *pujarinis* took good care of the deities: the deities were like people who slept, were fed, washed and dressed in clothes appropriate to the weather, including sweaters to combat the morning chill.

The *pujari* left a tray of food before the altar and returned to the kitchen, rang the bell another time, crossed the veil and took the food. Then, this food was returned to the pan, consecrating the entire meal.

An anthropologist observed the devotion to the divine personality and consciousness, materialised in the statue, called deity, which had a name (Sri Gauranga) and, at the same time, was also the Supreme God, Krishna. The deity Gauranga was a person, the very personality of Sri Chaitanya Mahaprabhu,[1] who was, according to the vaishnava conception, one incarnation of Krishna and Radha. Radha, the Divine Consort of Krishna, is the very materialisation of Krishna's feminine energy (*shakti*). Such a conception, simultaneously monist and dualist, seemed very strange – in the sense of instigating the reflection on the non-opposition of multiplicity and unity manifested in vaishnava ontology. It also challenged the impersonalist ideas that the anthropologist studied for many years.

If the divinity is a person, in contrast with the understandings of Buddhism or the non-dualist Vedanta of Shankara, how does one decipher the notion of the divine as a person simultaneously multiple and one? Different and identical in relation to humans?

A conceptual transformation, a thought that made a difference, was embodied. There, the anthropologist who wrote these lines, expressing the profoundly distanced language of the underlying narrator, felt the need for referring to herself with sonorous "I" and "Me". Nevertheless, the disquieting epistemological question remained: Is there a cognizant subject separate from the objective perceptual alterity? The restlessness was due to the intricacies of the notion of personhood in Vaishnavism and the implosion of the Euro-American notion of the person as an individual (Mauss 1985; Dumont 1986), degraded by the individualistic disconnection of the late neoliberal capitalism.

The question above corresponds to the phenomenological doubt regarding the possibility of the simultaneous unity and duality between the person and the world. The vaishnava notion of the person suggested such a possibility. And it also implied other theoretically relevant issues, to which I will return soon. My task was to suspend judgement, look for phenomenological reduction of the notions and overcome any rigid dichotomies that stunts the process of thought.

The environment of the ecovillage *Vrinda Bhumi* was propitious to knowledge and dialogue.

I asked Gopinatha, who coordinated the volunteering activities, for an interview with Puri Maharaj, the *sannyasin* who was the spiritual leader of *Vrinda Bhumi*. And through our dialogue the vaishnava notion of the person became clearer.

Our meeting happened after the morning meal. Maharaj talked to another volunteer and, later, asked her to call me. He was waiting at a wooden little table near the lodge. Maharaj was 52 years old. Almost 2 metres of a healthy white body dressed in saffron. He appeared to be distant, rigid, jovial and receptive at the same time. He did not seem ethereal, even though the fine cloth of his apparel expressed soft movements.

I mentioned some ontological issues and asked a question regarding the concept of the relation between matter (*prakriti*) and Consciousness (*Purusha*). After listening attentively, Maharaj answered:

– *Purusha* means enjoyer. We believe there is only one *Purusha*, who is Krishna. God is the Being who can enjoy everything. And we are part of this Being. There is this illusory energy, *maya-shakti*, and the material reality, *prakriti*. And there is also *jiva-shakti*, [the energy of] the individual soul. This individual being who is the soul tends to enjoy *prakriti* and *maya-shakti*. And it causes suffering, attachment. Every time one enjoys the material world, one creates a very strong attachment because one creates *karma* and a reactionary plane. One must pay for the enjoyment. It is like going to the market and taking many products, putting them in the cart. When one will leave, one must pay for everything. Then everything one enjoys in this material world; one may pay for the consequence of the enjoyment. Why? Because we are not *purushas*. *Purusha* is Krishna. He is the enjoyer. He is the energetic and we are the energy. We must enjoy what God has enjoyed first. This is the idea. And then one does not need to suffer the consequence of the karmic attachment. Often the soul is also considered *Purusha*, but only in the context where the soul can enjoy bliss. But *Purusha* is only one, the one who enjoys.

According to one of the most important books for every Vedic philosophy, the *Bhagavad Gita*, the material reality or Nature (*prakriti*) is the Field known by the *Purusha.*[2] Krishna is the *Purusha*, the Knower of the Field. He knows and enjoys the world through people – and this is the notion which leads to Krishna consciousness. Humans can purify their souls if they dedicate their senses to the enjoyment of the Observer, and then establish a kind of personal relation with the Universal Soul, this superior consciousness. Such a relation leads the individual soul to the liberation from the cycles in which the soul is tied. It is also a path to liberation from the illusion in which the soul is attached when the soul has not yet perceived its unity with the Universal Soul and the unity of the individual consciousness with the Universal Consciousness.

In the 13th chapter of Bhagavad Gita, Krishna said: "He who thus knoweth Spirit and Matter with its qualities, in whatsoever condition he may be, he shall not be born again". According to Maharaj:

– Krishna acts through us, but he does not participate in the Field. The Knower of the Field is the soul. The body is the Field. A person has the senses, mind, intelligence, and the false ego. These elements participate in all activities. Vedic theology postulates that *prakriti* is Krishna's servant, who takes care of living entities in this material world. *Maya shakti* [illusory energy] is like the prison warden. She keeps you in the jail. Krishna does not participate in this directly because he is on the spiritual plane. His participation is his expansion as *Paramatman* [great soul], like a radar in the heart of every living entity, like the voice of the consciousness. An example: On the branch of a tree, there are two birds. A bird eats one fruit, and this bird is the living entity enjoying the material world. And the other bird is this aspect of God, *Paramatman*, who is only observing. *Paramatman* delivers to us the results of our desires. If one has many desires, *Paramatman* can satisfy all these desires. He gives many bodies for the soul to satisfy the desires. Nevertheless, he is observing the soul. And when the soul looks at him, the soul stops enjoying because it understands that everything is to God. When we eat, we offer our food to God, and we wait for him to eat. And God eats with his transcendental and absolute senses. So, he can eat with his eyes and without even touching. What we eat is the remnant. *Prasada* means mercy. We are eating something that God ate first. Our aim is not stopping feeling or using our senses. We keep using our senses, purifying them. Our aim is to serve the Lord of the Senses, Rishikeshi. We serve the Lord of the Senses with our own senses. We use the senses to chant, listen, eat and by doing this we can satisfy God.

Now it is probably clear why the deity eats the offered food. If the energy of Krishna is manifested in the statue, then the statue is an aspect of Krishna who experiences the material reality through the bodies (Field). Krishna is the one who is being fed. Even though he does not participate directly in the activity of eating, he is satisfied through the food offered to him through the Fields that he, as the Knower of the Field, roams. The spiritual process (*sadhana*) – which involves learning, devotional service and chanting of the saint names – awakens in the incarnated person the comprehension of the inconceivable simultaneous unity and duality (*achintya-bheda-abheda*) between the Krishna Consciousness and the individual consciousness, and between the material reality/Nature (*prakriti*) and the Consciousness of the Enjoyer (*Purusha*). Devotional service is offering the results of all actions to Krishna in such a way that the person can act in the world, accomplishing their tasks, without attaching to the world. The Yoga of Devotion (*Bhakti Yoga*) dissolves the karmic effects one needs to face every time one enjoys the material reality.

The material reality is understood as an energy emanating from Krishna. And this energy is also consciousness, and such a consciousness is personal. Material reality is not the substance, but rather the shadow of it. According to Vaishnavism, the true substance of the world would be the personal Consciousness of Krishna.[3] The devotion to Krishna Consciousness implies the perception of the sacred in all extension of time and space. This spiritual experience creates the connection of the individual soul with the Universal Soul and does not dissolve the personal spiritual ego through final unification or annihilation (sometimes understood as Nirvana in Buddhism). To eliminate the individuality would prevent the person from serving Krishna – who is a person too – by means of intramundane action through the Field, which is the body, the matter and nature.

On the following day, Maharaj and I conversed, for an hour and a half. At the same place, after the morning meal, I explained I was not observing everything in the ecovillage because my research was delimited: I wanted to understand how one perceives the relation between person and world, person and nature. I asked him how he decided to become a *sannyasin.*

– My history is not spectacular, but a common history instead. I was always a common person who saw the material world as something to be enjoyed. But inside of me there was restlessness. I did not find a meaning in keeping working in the material world. I thought everything was futile because we lose everything when we die. I was a physical education teacher and I worked at the gym. We took care of our external beauty which is not important. It can be considered an almost useless profession, if we understand that, like Krishna told to Arjuna in the beginning of *Bhagavad Gita*, we are not the body, we are soul, and soul does not die. I was worried about the body. The first time I chanted the *Maha-Mantra Hare Krishna*, I awoke. It was the feeling of finding something I had already done. Something good and familiar. I felt that singing the mantra I could reach happiness. And it brought me hopefulness. It is a thing that depends on me. Sitting and chanting, practicing meditation. I did not need to look after anything anymore. I understood that happiness is not in things, but in being, sitting and meditating. I thought I needed to have things, before. If I wanted to eat something and satisfy my senses, I did. If I wanted to meet someone and satisfy my senses, feel happy, I did. If I wanted to buy a house, have a job and money… Then I saw it did not make sense. The meaning of life is being happy with oneself.

– What is the name of your guru? When were you initiated? I asked.

– I was initiated in India. His name was Srila Bhakti Pramode Puri Goswami Maharaj (1898–1999). He was 102 years old when he initiated me. A very pure person. A whole life of complete spiritual dedication. I passed through the first initiation (*Hare Rama*) with the mantra Hare Krishna and received some mantras to chant during the second initiation. That was my connection.

Puri Maharaj, like many people I met during my ethnographic travels, expressed dissatisfaction with the common way of living. He, like many people who dedicate their lives to the spiritual experience, did not see meaning in the hegemonic lifestyles. At a certain point of his trajectory, Maharaj chose to live a different experience, in connection with earth and spirituality. What did Maharaj understand for spirituality? The first answer was succinct: spirituality is the search for connection with God.

I insisted: – Spirituality is a word, a noun referring to an idea, a practice, a technique? And Maharaj replied with a broader explanation.

– Spirituality is related to movement. The movement of the finite in relation to the infinite. When the finite wants to connect with the infinite, this is spirituality. Searching this connection again, resume this connection, resume the identity – this is spirituality. Spirituality means dissolving the ego. The ego is this false conception one has about oneself: I am a man, I am a woman, I am white, I am black, I am old, I am young, I am beautiful, I am ugly. All of this is ego. And ego moves us away from spirituality. When one has ego, one does not have spirituality because this relation is inversely proportional. Ego works with concepts of "I" and "mine". And spirituality work with concepts of surrender, concepts of a new identity – this is spirituality. It is different from the selfish concept. It is the dissolution of the false identity and creation of a new identity.

– And how does one define this new identity? I asked.

– This new identity will be defined after a process. When one begins the spiritual practice, one achieves a devotee's body *sadhaka-deha*, a spiritualised material body. The devotee, through initiation, chants the *Maha-Mantra Hare Krishna*, but the body is a material body. Nevertheless, a material body which is already spiritualised. Then, this body enables one in the process of *Bhakti Yoga.* One reaches a state of devotion after many stages, since the phase of practice (*bajana-kryia*); passing by the phase of mitigation of the impurities (*anarta-nivritti*) – the phase in which everything undesirable is removed –; achieving the phase of stability where one develops a *roti*, a taste; until the moment the devotee begins to develop the perfect body, the body in the spiritual world (*siddha-deha*).

According to this interpretation of Vaishnavism, spirituality concerns a movement of the finite in relation to the infinite, a movement of the part connecting with the whole. The spiritual process of *Bhakti Yoga* – as the words of Maharaj made clear – involves the connection of the person to the Consciousness of Krishna, through a series of stages whose aim is the "creation of the I" (Mauss 1985: 13). Nevertheless, the fabrication of this vaishnava person is a process of dissolution of a false ego (*ahamkara*), and the creation of a new identity of the me. This vaishnava notion of the person, derived from the Vedas, also expresses one premise of *Samkhya*, which is, according to Mauss, "the school in which in point of fact must have preceded Buddhism" and "maintains the composite character of things and minds" (ibid.).

The process of composition of the vaishnava person is based on the connection with the divinity, by means of the devotional practice, and leads the person to the spiritualisation of the body as the Field through which Krishna knows and enjoys the material reality. Spirituality, understood as the connection between the personal consciousness and the Krishna Consciousness, implies the understanding of a previous connection between the personal consciousness and the matter, both external energies emanated from the all-pervasive aspect (*Brahman*) of the Supreme Personality of Godhead. Thus, such a conception encompasses the connection with nature, i.e., the connection with the material reality in which all organisms live, including the incarnated humans with their anthropomorphic bodies.

This idea of connection is related to the concept of *virat-rupa* (Universal Form of God). The Universal Form of God is the manifestation of Krishna. Maharaj emphasises that "everything is energy of God", from stones to trees. The Universal Form encompasses the multiple things one can admire in the world: mountains, seas and so forth. In sum, nature. *Virat-rupa* is the divine exteriorisation that happens through nature (*prakriti*). The anthropomorphic form of Krishna (his personal aspect, *Bhagavan*) is a manifestation of his personality as nature is the manifestation of his own energy (which is the aspect *Brahman*). Both manifestations are emanations of the Consciousness of Krishna. The *Brahman* aspect is like the sun that penetrates everything.

The search for connection which is spirituality is evidently present in the ontological conceptualisation of Vaishnavism. This search for connection shifts one's perception of life, of the body and existence. I asked Puri Maharaj: – Does spirituality shift the perception of nature? And he returned:

– Yes, I think that people become more sensitive, more contemplative of nature. There is respect. The devotee sees the trees as citizens because the tree has its soul. One must not cut it unnecessarily. The devotee sees the personality of the water. There is a deity who guides the water, *Varuna*. There is a deity who guides the air, *Vayu*. Rain is guided by *Indra*. *Surya* guides the sun. According to the Vedic culture, everything has a personality. All beings are people. And the devotee respects all these aspects of nature. This connection is developed more and more along the devotional practice. Each thing, animal, being, leave, little flower is seen as a part of the Supreme.

The beings of nature, these other beings, are considered people in the sense that they have their forms of soul and consciousness. Such a conception implies the notion of metempsychosis, according to which the transformation of life forms and their inherent consciousness is not a unilinear evolutionary process. The process is cyclical. The personal consciousness that animates the human body can animate the body of a tree. After death, the consciousness can animate bodies of the animal kind.

The notion of other beings and things in nature as people reminds me of animist conceptions. Animism refers to the idea that every being possesses a similar interiority, although with multiple physicalities; there is a continuity of the souls and discontinuity of the bodies (Descola 2006). According to the Vedas, there are 8,400,000 life forms, and 400,000 are human forms. The notion of the same substance infused with consciousness reminds me of the monadological concept of discontinuity of phenomena and continuity of the substance. Despite their physical discontinuity, the phenomena called "Field" share this same ontological substance, which is *prakriti* – matter/energy – and *prakriti* is infused with consciousness. Vaishnavism introduces a monist concept of the relation between mind and nature.[4] It introduces a monist thought where energy is to matter what Consciousness is to energy (the homology being **matter:: energy:: Consciousness**), notwithstanding being dualistic in how it frames the relationship between individual consciousness and Krishna Consciousness.

I want to emphasise this shared premise of spirituality and ecological thought: the existence of a "pattern which connects" all living creatures (Bateson 1979). Such a notion of connection – of which Vaishnavism as epistemology is an example – underlies both ecological thought and spirituality. Both rely on the search for the observation of relations and connections between beings and things in the world; the observation of the "sacred unity of the biosphere" (Bateson 1979). Both conceive the continuity between matter and consciousness or even assert that matter is consciousness, overcoming the Cartesian prejudice on the inert matter. The argument of my dissertation becomes even more clear – spirituality and ecological thought are the foundations of a sacred perception of the environment:

> To perceive the sacredness of the environment, things or situations of daily life implies a specific modality of seeing and acting in the world. In a sense, the sacred does not refer to the things by themselves, but it rather refers to this singular way of seeing (or reading) things beyond what is commonly seen.
>
> (Chiesa 2017: 428)

At five a.m., the *pujari* blew a big shell, generating a sound that served as an alarm. The *arati* (offering rite) would begin. The sound of the shell translated the abundance of *Lakshmi*, who invited people to the temple. The *pujari* opened the veil of the altar and rang a little bell. The devotees arrived. Women sat on the right side. Men on the left. Someone sang the first verses of the chanting offered to Krishna and Radha.

The *sankirtana*, the congregational chanting, included prayers, greetings to the masters, preceptor gurus and deities, as well as chanting in Sanskrit, Hindi, Bengali and a few in Portuguese too. The small candlestick with four candles made of *ghee* was taken by a devotee. Each person, when their own turn came, raised the left hand, almost touching the flame, and with the warmth of the fire on the hand, touched the forehead and the cardiac region.

Figure 4.3 Feminine *ashram*

Later, there was prayer and chanting to the sacred plant *tulasi*, which was placed in the centre of the temple. This plant was very relevant for the devotees. The symbols in the foreheads of the monks were the V of Vishnu and the flower of *tulasi*. Men circumnavigated the plant, three times. And then women, three times. Finally, the devotees chanted some verses whilst everybody circumnavigated around the *tulasi*. The last prayer blessed the material bodies, so that the devotion to Radha and Krishna would be their highest pleasure, like the nectar of the white lotus flower, the transcendental love.

The religious meaning in vaishnava philosophy is the reconnection of the individual soul (*jiva*) with the Universal Soul (*Krishna*) through devotional service founded in *Bhakti Yoga*. Yoga means connection, union. Vaishnavism as a religion provides the person with techniques for reconnecting with the Supreme Personality of Godhead through action in the world. After reconnecting, *jiva* is liberated from the cycle of births and deaths (*samsaras*), but *jiva*, as an individual personality, remains the same, being simultaneously different and identical to Krishna.

According to such a conception, when the individual consciousness/soul leaves the physical bodies it used along its rebirths, it no longer shares the material energy (*para-prakriti*) of the beings and things in the world, but the

Figure 4.4 Statue of Krishna subjugating the serpent Kalaya, with the lodge behind

spiritualised energy (*apara-prakriti*) remains. The soul then lives in the spiritual region, although it does not become *Purusha*, which is singular and contain all the souls, even though it is not contained by the souls. Vaishnavism conceives the unity of nature and people and their duality in relation to the divinity. And, simultaneously, it also conceives the material reality of humans and non-humans as united with *Purusha*, the Consciousness behind it.

Vaishnavism is also a practical philosophy. Bakhti Devi Dasi, for example, considered art to be her devotional service. All the statues were made by Bhakti, an Argentinian devotee who dwelt in the ecovillage. She was 31 years old. Bakhti was born in Buenos Aires and lived there until she was 27. She had lived comfortably, but she decided to leave everything behind, detach, and left, looking for herself. Her art allowed her to be in the constant contact with the earth and other people. With her foreign accent, Bhakti told me how important the mud was in her life. The wet earth helped her to feel rooted because she had an inner flow that propelled her to this movement of search. When I met her, Bhakti was building flower beds, and she felt that the ecovillage was her place.

When Bhakti left Buenos Aires, she travelled in Argentina and crossed the frontier. She was hitchhiking and someone robbed her. She lost everything.

Money. Art. She was left on the road with the clothes she was wearing. Absolutely alone. She told me she did not know Portuguese, and so she surrendered to the cosmos. And everything she needed, she received. She camped alone on a deserted beach for six months. She felt she needed to explore alternative communities and decided to search for it. According to her, this movement of search is a particularly profound spiritual experience, which words could not express.

Finally, Bakhti discovered *Vrinda Bhumi*, without any planning. There were many alternative communities in the region where the ecovillage was situated. To Bhakti, this region had not been infected by capitalism, being extremely isolated within a chain of mountains, Serra da Mantiqueira. The water was abundant and pure. The place was propitious to the spiritual practice of connection with the earth. Bhakti said: "There is something beyond what is perceived in this place, an energetic force that does not exist elsewhere, but I do not know why". The place was like an "energetic lung", she defined.

The person, the flesh of the world, the lived world and the social world

It was a dry and cold day of July. The routine for devotees and volunteers in *Vrinda Bhumi* was disciplined. The tasks for volunteering included the management of earth, seeds and seedlings, apiculture, planting, fabrication of adobe bricks and cleaning activities.

At 7 a.m., Gopinatha rang the bell in the communal space above the lodge. This building was being bioconstructed using the cob wall. The volunteers awoke and helped with cleaning and organisation of the common space, situated above the lodge. Some volunteers took firewood for the kitchen. Others took cow excrements to be used in the processes of bioconstruction. Then, the volunteers could attend the daily classes on philosophical themes, led by the *pujari* or *pujarini*. The classes took place at 8 a.m. Later, the first meal was served. Everybody worked from 11 a.m. to 3 p.m. Volunteers had free time until 6 p.m., when the second and last meal was served.

Most classes offered in the period of July 2018 concerned Vedic astrology. Often people created comparative explanations in relation to the Western astrology. On July 27, when the moon reached her zenith, a lunar eclipse also occurred. Three celestial events aligned on the same day: the supermoon, the blood moon and the lunar eclipse, when the earth is between the moon and the sun.

Chaitanya, a *pujari*, led us to the lawn, outside the ecovillage. We sat on the dewed ground. The fresh wind contrasted with the sun's heat on the skin. Chaitanya explained the possible malefic effects of the planets Rahu and Ketu during the eclipse, and for that reason, he taught us mantras that, somehow, would cancel those effects and ease the influence of the stars.

The mantras were conceived as "spiritual technology". Each mantra was related to a *yantra* (a graphic tool), *mudras* (gestures) and *karratis* (energetic protection shields). The word mantra is composed of two particles: *man* – mind or thought; and *tra* – instrumentality. Mantra is an instrument for controlling and directing the attention of the mind.

The sound vibrations of each mantra had different aims. The mantra *Om chandraya namaha*, for example, enabled the connection with lunar energy. The mantra *Om shanti* created a peaceful atmosphere. The Great Mantra of liberation, *Maha-Mantra*, was considered the most relevant because it does not fit only one specific situation. Chaitanya explained, it had a broad effect on consciousness. When one chants Hare Krishna, one reaches a transcendental intelligence, "an accurate perception of this world". The *Maha-Mantra Hare Krishna* required neither *asanas* – the positions of *Hatha Yoga* – nor rites. This is the reason why it was considered adaptable to the common lifestyle of the devotees who lived in the cities. *Bhakti Yoga* was a devotional service which could be practised everywhere. Chaitanya said that *Yoga* does not mean "stopping the mind and staying still", but rather acting in the world, thinking of Krishna while doing the activities.

Bhakti Yoga can be practised by anyone. Even though we were at this ecovillage, which was also a monastery, vaishnavas circulated around the world: they were at the universities, streets, working, making their art. The spiritual practice of the devotees, and the inner cultivation of Krishna Consciousness, was not opposed to the world. *Bhakti Yoga* was associated with the movement and did not require renouncing daily life. The action of the devotee was also purifying. The action in the world was not a "profanation" of the spiritual practice. Regardless, sooner or later, the inner world would be externalised in concrete action.

The same morning, I had the opportunity to speak with Prema Devi Dasi. Her name meant "servant of the spontaneous love of the Goddess". Prema was 28 years old when I met her. She was born in Rio de Janeiro and studied Administration. She was the *pujarini* in the temple and had a strong connection with a feminine spirituality, profoundly tied to Earth (*Bhumi*). Prema understood spirituality as the development of sensitivity to the environment, searching for the connection with the feminine characteristics of nature, such as care. Prema said: "When one develops maternal characteristics, one begins to want to take care of everybody. One looks at others with tolerance and tries to learn and teach, supporting the growth of others". Prema, like many people with whom I had dialogue, was deeply disappointed with the common way of living. During a backpack travel, she discovered Vaishnavism and a new way of living. She showed me how environmental activism is present in the vaishnava thought of Vrinda institution. The leader Vrinda is Swami Paramadvaiti, one of the activists who founded the World Conscious Pact, which galvanises many webs of different activisms, such as Indigenous rights,

Mother Earth rights, food justice, animal liberation, land reform, biodiversity and protection of seeds and water. The project had organised many Kivas, ceremonies (cosmo)political and spiritual at once, led by different Indigenous peoples of Americas. An important name of World Conscious Pact is Dr. Vandana Shiva – physicist, philosopher, environmentalist and ecofeminist – who has defended the Pacific boycott of multinational corporations by means of a conscious consumption and refusal of industrial food.[5]

Prema, replying to my question on spiritual experience being the source for activism, said:

– When I first got to the temple, I learnt that it was possible to change the world through the transformation of the consciousness. It is very important to begin from within, inside one's own world and relations. I got here willing to change the world. I raised flags, revolted by the conditions of the world. But my own daily relations were conflicting with my own position. I think it is possible to change the world if one is an example to be followed. And it is very difficult to change this world because every time one engages with it, the more they see the dirt. The spiritual path is a form of purification. One sees, recognises, cleanses oneself and walks again. It is strong but it is what really transforms the world of someone, the world with which one relates. Of course, it can transform the world, especially if many people achieve this shift of perception. This is our Guru's main goal. He is a true activist. He galvanises people who want to change the world and connect them to the service. When one changes their consciousness, one changes attitudes, from consumption to personal relationships. This a chain effect. Sometimes, speaking out is not necessary to be an activist because life itself is activism. More than speaking out is being. Through example, one inspires change.

According to Prema, the spiritual path leads to the transformation of the personal world. A process of seeing the dirt in this personal world, in order to clean it. Through the comprehension and change of this personal world it would be possible to change the larger world that encompasses the personal world. The number of people who realise this inner work influences the world, generating a chain effect. Nevertheless, the change is also qualitative in character. I do not understand this notion as simple addition of transformed personal worlds that result in the ultimate transformation of the larger world.

The issue is inquiring how – to use a post-structuralist term – this "relation" between person and the world happens. And what is the "engagement" of the person in the world (Ingold 2005).

In his *Cartesian Meditations*, Husserl suggested that the world exists to the extent that it exists to someone. The meaning of the world is in the person who perceives the world, and to whom the world exists. The perceived world appears as a "unity of meaning" that transcends the person. Each person has a specific experience of the world, and this experience can be understood as the life-world. The experienced world is, thus, the lived world. This is the reason why each person "has his experiences, his appearances and appearance-unities, his world-phenomenon; whereas the experienced world exists in itself, over against all experiencing subjects and their world-phenomena" (Husserl 1960: 91).

The ideas of Prema point exactly to this alterity between what can be called "inner world" and "external world". The "external world" is a larger world in which the person experiences their own "inner world". Therefore, the experience of spirituality can be understood as the transformation of the "inner world" and the "external world". How does the transformation of the "inner world" reflect on the "external world"?

There is a false dichotomy that could be associated with the two worlds. Hence, I write "worlds" in quotation marks. The "external world" engulfs the "inner world" in such a way they are interwoven. One is the reflection of the other; they resonate with the same pitch. One cannot exist without the other, even if it is understood that the external world is already there long before I begin to think on the possibility of knowing it. Both are interwoven, even if I understand that the meaning of the world is nothing but the meaning that is extracted from our experiences, thoughts, value judgements and actions (Husserl 1960). The "inner world" is, nothing less, the world experienced by the person. And this world often appears as something external. The "inner world" is the very person who perceives and knows the "external world". The whole lived experience is the very person; it is the lived world.

Merleau-Ponty also discovered the same indivisibility of person and world:

> The world is inseparable from the subject, but from a subject which is nothing but a project of the world, and the subject is inseparable from the world, but from a world which the subject itself projects. The subject is a being-in-the-world and the world remains "subjective" since its texture and articulations are traced out by the subject's movement of transcendence.
> (Merleau-Ponty 2005: 499–500)

The world exists because there is someone who perceives it. And what can be known in the world are subjective projections of the perceiver, who, being different from the world, lives in a world that cannot be nothing but "their" world. The world which seems external is inseparable from the person's inner world.[6]

> According to Merleau-Ponty, the world sustains the subject's body and moves with it, delimiting its field of experiential and perceptual exploration. As a bodily condition of the subject, the world is experienced as a

> constitutive element of the subject-body that inhabits the world, and not only as an objective reference external to the subjects that moves in it.
> (Carvalho and Steil 2008: 292)

The lesson one learns at this point is that the sensible world is in the sensitive person and the sensitive person is in the sensible world – from such an entwining rises the notion that "we are the world that thinks itself" and "that the world is at the heart of our flesh" (Merleau-Ponty 1968 :136). Thus, the duality between internal and external is dissolved. The flesh of the person branches out into the world as the world spreads within the flesh of the person. The inside of the person corresponds to the outside of the world and the inside of the world corresponds to the outside of the person. The body of the perceiver is made of the same flesh of the world. Nevertheless, instead of understanding such a relation in terms of a body-world relation, the empirical example that elicited all these lines seems to enable only a notion of the "person-world". The person is constituted and transformed in the flesh of the world. And if the world is made flesh through lived experience, the world becomes the world of the person, the person-world, which is the experienced world, the lived world.

"Where are we to put the limit between the body and the world, since the world is flesh?" (Merleau-Ponty 1968: 138). In order to describe the "reciprocal insertion and intertwining" of body and world, Merleau-Ponty presents the figure of two concentric circles. The two circles show that the external world and the inner world are inextricably entwined in such a way that, although being different, they are the same circle. I understand that the notion of the flesh of the world refers to the interstitial space that entwines the world and the skin of the person. Drawing inspiration from Merleau-Ponty, I created an analogy between the terms body and world in relation to inner world and external world. Due to the ethnographic reference registered in the ecovillage *Vrinda Bhumi*, I propose an alternative figure to express this idea. The symbol of yin-yang. The black dot is the external word within the white portion, which is the inner world. And the white dot is the inner world within the black portion, the external world.

How does self-transformation reflect on the flesh of the world? The concept of experiencing transformation of the lived world has not been sufficient to understand how this self-transformation is embodied. A post-structuralist notion can be useful, if it is approached carefully.

The question above poses a problem. Spirituality – understood as the process from which results a different sensibility regarding the person with the world and perception of the world – transforms one's lived world, which can only be a social world. This shift of perception is manifested in the social relations in which the person is entangled. This idea will make sense if we accept that the relational matrix of social life is not beyond people – this is the concept Strathern (1992) learnt from Melanesians. It makes sense if we bear in mind that "people are subverted by the actions of other people" (Strathern 1988: 102). Therefore, it is easy to admit that social relations which compose the person virtually contain all the potency of a broader change, and if such a change does not happen by causality, social relations carry the latency of an idea that becomes flesh through the contiguity of people who share it.[7]

The problem is that if social relations constitute the person, the person contains not only the possibility of a broader change, but many other numberless "possibles" (Tarde 1874). The "possibles" can be understood as the multiple virtualities not yet actualised. Notwithstanding the fact that the virtualities were not yet actualised, it does mean they are not real.

> In principle, every reality is conceived as containing essentially an excess of potency over the act. And it is exactly this excess of potency over the act that constitutes, to my eyes, the set of non-realised possibles or, if you will, conditional certainties. Consequently, the possible is an intimate part of the real, even though it is not the real: and these two terms are reciprocal. The intelligence of the facts requires, therefore, knowing the possible.
>
> (Tarde 1874: 14–15)

If it is possible that change happens by the sharing of ideas and examples of change through social relations, then one's inner change is, per se, social. One's inner change as social transformation is only possible because the very person – composed by social relations – is society. Maybe there is fecundity in the idea that "everything is a society, that every phenomenon is a social fact" (Tarde 2012: 28). Although it seems poetic, I note that if every particle is an association of infinitesimal particles, each person is a composite particle of different social relations and each person is like a constellation of changing particles orbiting around the sun.

Nevertheless, bearing in mind the reflections of the previous chapter, this relational matrix of human life, where social worlds are lived by people, may be understood as an entanglement, and not as a relation between individual and society or human and non-human as two separate ontological realms. If all beings live entangled with the earth, it is impossible to say that we live between the sky and earth, but rather in between the sky and the earth (Ingold 2015). We live in this indetermined and transformative space which continuously entwines humans and nature through flows of water, plants, seeds exchanged by human hands, seeds taken by the birds, precipitation, rainwater harvesting, burning of fossil fuels and atmospheric gases. The flesh of the world is not simply a relation between person and world, between the skin of the person who perceives the environment and the perceptual *medium* against which the body moves. The flesh of the world is also manifested in this space which is in between. The flesh of the world is the very life world, which is nothing but the social world entangling environment and people in continuous movement. Indeed, the image of *yin-yang* materialises such an entanglement very well.

Notes

1 Chaitanya Mahaprabhu (1486–1533) catalysed the philosophical and religious movement called *Gaudya Vaishanavismo* (Adami 2012). He was responsible for the popularisation of the *Maha-Mantra* Hare Krishna through his practice of devotional chanting in public spaces, galvanising people of different castes. "The expansion of the identity of Chaitanya representing a broader expression of Vishnu himself and the recognition of Chaitanya as a dual incarnation of Krishna and Radha are the root-foundations of the tradition Gaudiya Vaishnava" (ibid.: 87).

2 The narrative of the *Bhagavad Gita* consists of the final part of the epic poem *Mahābhārata,* which is attributed to the erudite Vyasadeva (Century III BCE?). It registers the dialogue between Krishna (the Supreme Personality of Godhead) and Arjuna, a member of the warrior caste (*Kshatriya*). Arjuna is about to begin a battle with his malefic relatives and, even though he knows the action is necessary, he is hesitating. All the dialogue points to the concept of the action conscious of Krishna, which does not cause karmic effects that would exist otherwise.

3 The conscious aspect of the ontological substance of the world reminds me of the conception of substance according to Heidegger. "By substance we can understand nothing else than an entity which is in such a way that it needs no other entity in order to be. The Being of a 'substance' is characterized by not needing anything" (1962: 125). Gaudya Vaishnavism asserts that Krishna is independent from the world, although he is the matter from which all beings and things are made. In the *Bhagavad* Gita, seventh chapter, Krishna defines himself: "Earth, water, fire, air, ether, Mind and Reason also and Egoism – these are the eightfold division of My nature". And then: "The pure fragrance of earth and the brilliance in fire am I: the life in all beings am I, and the austerity in ascetics".

4 I understand that such a conception, apropos, can be linked to the concepts of *Pleroma* and *Creatura,* the two aspects of the real suggested by Bateson (1972). *Pleroma* regards the world of the events caused by forces and impacts, the common material universe; *Creatura* regards the world of information, ideas and differences (Bateson 1972, 1979). Nevertheless, this dyad cannot be considered a mere substitute for the Cartesian dichotomy of mind and nature. Conversely, "we cannot see *Pleroma and Creatura* as two separate dimensions, but rather two combined levels. Everything in *Creatura* also exists in *Pleroma* and the later depends on the former to be accessed" (Chiesa 2017: 422). These two levels share the same substance and, therefore, *Pleroma* can be known through *Creatura,* even though Pleroma cannot be known *per* se, like a territory can be known through a map, even if the map is not the territory. I would add another dyad: the embodied human mind and the world, which is also mind. The microcosmic mind is intricately related to the other, a microcosmic mind: the mind-world and the mind-body are two terms of the same relationship. Even though they work in different scales, their relation is monist. Maybe, the proposition of dyads is just a way towards thought organisation.

5 The Project website is available at https://worldconsciouspact.org. Regarding Vandana Shiva, she has written many books, including *Monocultures of the Mind* (1993).

6 I want to emphasise that I am not arguing that each person "constructs" a particular world inside the head, as a representation of a single world which is interpreted in multiple ways. The idea of world is beyond the dichotomy of nature and culture,

the opposition between realism and idealism, the relativism and universalism and beyond the representational perspective. The idea of world here refers to the lived world, the environment in which living organisms inhabit and move. The perception constitutes the person's inner world because what one perceives, by perceiving the world, is oneself. "'What is the product of perception?'" Ingold (2005: 95) asked. And the answer: 'The perceiver'. And it happens because perception "involves the whole person, in an active engagement with his or her environment" (ibid). Perception happens just because the person has the perception of oneself in the world, and not because the world is perceived by a separate mind. Even if the "external world" presents itself as transcendental.

7 Maybe I am dealing with a "flat ontology" (De Landa 2002), according to which society is not understood as a totality, but rather as a dynamic whole that emerges from the interaction between singular entities. It is important to emphasise that the possible (virtual) relations are not opposed to the real, but to the actual (Escobar 2007). I understand that it is useful to retain the conceptualisation of the possible and the real shared by Tarde and De Landa without immediately engaging with the Deleuzian notion of reality, which encompasses three ontological dimensions: the virtual, the intensive and the actual. It is important to note that the main path that led me to the possible and to flat ontology were my post-structuralist readings, instead of the principle of symmetry.

References

Adami, V. H. 2012. "Modelos e Moldes de tradições: A hermenêutica do movimento Hare Krishna (ISKCON) sobre a tradição Gaudya Vaishnava". *Sacrilegens* 9(2): 86–104.

Bateson, G. 1972. *Steps to an ecology of mind*. New York: Ballantine Books.

——— 1979. *Mind and nature: A necessary unit*. New York: EP Dutton.

Carvalho, I. C. M., and Steil, C. A. 2008. "A sacralização da natureza e a 'naturalização' do sagrado: Aportes teóricos para a compreensão dos entrecruzamentos entre saúde, ecologia e espiritualidade". *Ambiente & Sociedade* 11(2): 289–305.

Chiesa, Gustavo Ruiz. 2017. "À procura da vida: Pensando com Gregory Bateson e Tim Ingold a respeito de uma percepção sagrada do ambiente". *Revista de Antropologia* 60(2): 410–435.

De Landa, M. 2002. *A new ontology for the social sciences*. Paper presented at the workshop New Ontologies: Transdisciplinary Objects, University of Illinois, 30 March, Mimeo.

Descola, P. 2006. "Beyond nature and culture". *Proceedings of the British Academy* 139: 137–155.

Dumont, L. 1986. *Essays on individualism: Modern ideology in anthropological perspective*. Chicago: University of Chicago Press.

Escobar, A. 2007. "The 'ontological turn' in social theory. A commentary on 'human geography without scale', by Sallie Marston, John Paul Jones II and Keith Woodward". *Transactions of the Institute of British Geographers* 32(1): 106–111.

Heidegger, M. 1962. *Being and time*. Trans. J. Macquarrie and Edward Robinson. Oxford: Blackwell.

Husserl, E. 1960. *Cartesian meditations*. Trans. Dorion Cairns. The Hague: Martinus Nijhoff Publishers.

Ingold, T. 2005. "Human worlds are culturally constructed". In *Key debates in anthropology*, ed. Tim Ingold. London: Routledge, pp. 112–118.

——— 2015. *The life of lines.* New York: Routledge.

Mauss, M. 1985. "A category of the human spirit: The notion of person; the notion of self". In *The category of the person*, eds. M. Carrithers, S. Collins and S. Lukes. Cambridge: Cambridge University Press, pp. 1–25.

Merleau-Ponty, M. 1968. *The visible and the invisible*. Evanston, IL: Northwestern University Press.

——— 2005. *Phenomenology of perception*. London & New York: Routledge.

Shiva, V. 1993. *Monocultures of the mind.* Penang: Third World Network.

Strathern, M. 1988. *The gender of the gift.* Berkeley: University of California Press.

——— 1992. *Reproducing the future: Essays on anthropology, kinship and the new reproductive technologies*. Manchester: Manchester University Press.

Tarde, G. 1874. *Les possibles: Fragment d'un ouvrage de jeneusse inédit.* Available in https://www.enap.justice.fr/sites/default/files/histoire_parcours1_tarde_biblio_les_possibles.pdf

——— 2012. *Monadology and sociology*. Melbourne, AU: re.press.

Denouement

Breaking through the social membrane

The three notions of spirituality that, recursively, were expressed by the people with whom I dialogued do not need to be necessarily considered separately. These spiritualities are formulated in slightly different terms, but each is related to ecological epistemologies that overcome the dichotomies of the modern Euro-American thought.

Nevertheless, these three notions of spirituality, explored alongside this research, share the same thread that inextricably entangles them. They regard experiences that cannot be reduced to the field of discursivity, of conceptualisation. As a matter of fact, the interlocutors were able to express these notions because such notions are rooted in lived experience. The knowledge one constructs of the world can only stem from the experience of inhabiting and knowing the world, simultaneously.

This interdependency of concept and experience led me to explore both post-structuralism and phenomenology. These lines of theoretical and methodological language provided me Ariadne's thread with which I could walk inside the research labyrinth, even though I had the impression that I was walking in circles sometimes. Or an analytical spiral where each time my mind seemed close to the centre, the core, my reflections were pushed away. In this labyrinth there was no centre, only an attentive path.

If I could present a formulation in which the three notions are integrated, I would say that spirituality can be understood as a transformative experience that engenders the perception of the intricate entanglement of the outside of the world and the inside of the person. Thus, the experience of spirituality transforms the perception of the person regarding their relations with the world.

Nature is not perceived as an external alterity, a perceptual horizon situated outside the limits of the perceiver's body, but rather as the environment in which the perceiver is continuously immersed. An environment which is, itself, continuously becoming part of the porous perceiver. Likewise, the social world in which the person lives is fundamentally the lived world whose trend is being perceived as a transcendental phenomenon. But society, the set

DOI: 10.4324/9781003378853-8

of all social relations, is immanent to the individual. Or more precisely, the life of a person is constituted as a social world, which is always entangled to other people's lifeworlds, as lifelines that correspond to each other's movements. The environment and the person are like the ocean and the water drop. Individual and society, or rather, person and world, are like similar sounds that vibrate in different octaves.

What we tend to perceive in society as an externalised phenomenon is, indeed, inside people. Individuals bring the seeds of collective behaviour. If society is nothing but the world lived by the person, all the problems and virtues of the lived social world are manifested through the most intimate relationships with others. And the ways humans relate to each other is extended to the environment. The entanglement of person, social world and environment is immanent. For this very reason, self-transformation is equivalent to the transformation of one's lived world, which is nothing but the transformation of the social world. I consider the lived social world as this environment – in which earth and all kinship of humans and non-humans – is experienced by people.

Following this monist notion, the modern national state, for example, could not be considered an external entity outside the daily practices of people who produce it. If a person shares on *Facebook* their position regarding the integration of Indigenous populations to the national society and the consequent expropriation of their ancient lands; or if a person expresses the opinion that agrobusiness creates jobs and leverages the national economy, despite soil desertification and the use of herbicides, that person actualises and embodies, at that moment, the colonial gesture of the developmental state. Ecological crisis as a "global" phenomenon is also the resonance of people's attitudes, such as the deposit of solid residues on an estuary of the Yangtze River in China or electronic waste on the river Arroio Dilúvio riverbed, in Porto Alegre. The old lesson "as above so below" implies considering the individual as an immanent correspondence of society. What one observes in the lived world is a complex set of personal discourses and gestures that surpasses the microcosmic scales and, at the same time, is reproduced through them.[1]

The monist notion of the lived social world includes the premise of fractality, considering that wholes are incommensurable with the sum of the parts because the parts are wholes (Strathern 1992). Ultimately, the person is per se, as a hologram, the whole of nature and life-world. Theoretically, all levels of analysis that social scientists tend to heuristically situate outside the person – such as personal, local, regional, national, global levels and so forth – are all entwined. Not as juxtapositions of pieces of reality, but rather as practices and discourses which are immanent, intrinsic to the person and situated in the lived social world.[2]

Hence, social relations are present in the gestures of people in their daily interactions, even those which are hidden or cannot be evidently observed. The gestures of people contain virtually all possible relations in their social worlds, since the parts are analogous to the whole. Despite being different, they are mutually implicated. Mining, for example, generates wide effects beyond the biodiversity and ecosystem of the exploited place. The latent (virtual) values of consumerism, egocentrism and competition that propel capitalist accumulation are embodied (actualised) through people's daily gestures, but they also engender a planetary social imbalance, which surpasses the sum of personal attitudes because the destructive effects of the exploitation of human life for monetary purposes reflect beyond any particular situation. Another example are the challenges faced by countries like Brazil in the context of the Covid-19 pandemic. Political leaders and other state agents have propagated science denialism and refused the health protocols publicised by the World Health Organization. Such an attitude was fed by supposedly individual choices of common people in their lives. Choices like refusing social distancing or avoiding face masks. The gesture of the citizens who denied the dangerous virus contamination, as well as the very existence of the pandemic, reflected the din of a public administration based on fake news, apology of ignorance and the spread of disinformation.

Again, following the same monist notion, the gum an adult man drops from the car window or the wires of my cellphone left on that shelf full of electronic waste are, despite their apparent dissimilarity, factors that contribute to the big problem of ecological crisis. However, the deforestation of Amazon is by no means comparable, in terms of scale, to the plastic bags used daily by one person.

Indeed, the notion of spirituality as transformation of perception of the environment and the consequent transformation of the lived social world is deeply tied to the meaning of an ecological epistemology that dissolves the dichotomy of person and world. Instead of one mere individual transformative experience, "ecological spiritualities are fabricating subjects projected in the world, in continuity with the rivers, mountains, oceans, atmosphere, trees", opening the perceptual horizon of "a subject projected and distributed in the world-environment" (Steil and Carvalho 2021: 10). The experience of such a modality of spirituality is not restricted to inner life, but rather inspires people for their actions in the lived social world.

Alongside the ethnographic narrative, one of my aims was to describe the knowledge lived in the ecovillages, without necessarily comparing such lived knowledge with other forms of knowing socially recognised. The epistemic approach aims to emphasise the epistemological validity of the observed culture because knowledge is also culture. Hence, the critical description of local epistemologies considered the relevance of the modalities of knowledge production in ecovillages.

On this epistemic level, the similarity of the four ecovillages is permaculture and its principles. Permaculture can be considered an ecological epistemology that, in many ways, is experienced in the four ecovillages. The ecological spiritualities enacted by those people are the experiential counterpart for the shared concepts relative to permaculture, agroforestry and bioconstruction.

In the ecovillages, there is an immanent spirituality experienced as connections with nature. It cannot be reduced to an individualistic concept or to a concept rooted in the dichotomy of spirit and matter.

The individualistic concept of spirituality, such as the notion explored by the research on the New Age, addresses the idea that the experience with the sacred happens only on the personal level. Elsewhere, I argued that the aprioristic association of spirituality and the New Age fails to grasp many important aspects, especially the political realm and the social action influenced by spirituality (Gonçalves Brito 2020). The image of the middle-class intellectualised person who meditates, buys therapeutic services, attends weekend workshops and circulates through the neo-esoteric circuits of the big cities is a very different phenomenon from the ecological spirituality in ecovillages.

I think that my present study also shows the methodological pitfalls of the research which tightly associates spirituality and the New Age. Even though such an association is perfectly adequate to certain empirical contexts and analysis, it obscures any comprehension of spirituality beyond the ideas that it is an almost hedonist cult of well-being of the autonomous self. The people whom I engaged in dialogue during my ethnographic travels could not be considered the kind of middle-class individuals who close their eyes to the hardships of the world whilst meditating in their comfortable apartment in the big cities.

I argue that all this methodological reasoning enables an investigation on spirituality beyond the assumptions that have been depleted by the previous studies. I suggest that spirituality is not the only phenomenon imbricated with the secular frame of modernity (Veer 2009) because the emergency of ecology can also be understood as a phenomenon resulting from the immanentisation of the experiential reality. This immanentisation regards the disjunction of state and religion and the modern premise of secularity, which has freed the minds from the loaded idea of faith being the only logic able to explain the human existential anxieties. Thus, there was a shift from a transcendent cosmological order to an immanent order in which nature, earth and all beings, always entwined in ecosystems, are sacralised. I call this specific phenomenon *immanent sacred of nature*.

Notwithstanding the fact that the two ecovillages I visited were bound to religious institutions, the proximity to natural environments propitiated another reality. The two vaishnava ecovillages were bound to different religious institutions, but people shared the same philosophy. The interest in the analysis of the category of religion – a Euro-American category which could not simply be applied to the Hindu or Vedic context (Asad 1993) – was shifted to the description of the experience of ecological spiritualities. Especially the three notions of spirituality repeated through the dialogues, discoveries, paths and ethnographic travels.

Alongside this study, similarities and differences between ecological spiritualities and New Age spiritualities were noticed. Nevertheless, my methodological choice has been to avoid a previous association of spirituality and the New Age, given the semantic load of the term New Age. Such a choice allows us to avoid certain theoretical pitfalls that seem pervasive in the studies of the New Age, especially the notion that spirituality is merely one more product in the religious market of self-satisfaction. That notion does not encompass the multiple experiences and what the keenest of my interlocutors understand as spirituality.

Spirituality is an experience. However, such an experience is not necessarily alienating, like many studies seem to suggest. Spirituality is a category that presents a multiplicity of meanings, depending on the concrete situation in which it appears. But it does not mean that all particular conceptions of spirituality are valid on the broader level of analysis because, in order to be considered an analytical category, the notion of spirituality must be observed in different places, as a relevant element in the social worlds.

I remember that, during our dialogue, Wilson of Project Muriqui Assu remarked that the hegemonic lifestyle – where people work long hours daily in a cycle of too much effort to achieve the little money that the capitalist system allows them to have – practically prevents the experience of spirituality. Spirituality, according to his conception, concerned this commitment with social change that emerged from the inner change of the individual. The paradox of this conception is that it is rooted in the personal and mystic experience of Wilson, who dedicated passionately to a lifelong intramundane experience of personal transformation, by means of his affiliation and active participation in alternative communities bound to Children of God/Family International.

One of the threads that runs through the conceptual core of this dissertation is the paradox of a spiritual experience that inexorably happens inside the person and, at the same time, overflows to the lived world.

The experience of people who live in ecovillages is very distinct from those who live in cities, not only in terms of physical space, but mainly in terms of personal and collective choices which cannot be entirely explained by financial conditions or sociological circumstances. Notwithstanding the fact that many among them grew up in middle-class families and studied in formal institutions of high education. Regardless of those people's material conditions of existence, everyone rebelled deeply against the bourgeois morality and the hegemonic political-economic system of many countries around planet Earth.

Theoretically, I have tried to contribute towards the flourishing of a monist notion of person and the world, considering the obsolescence of the concept of society. I join the social scientists that want not only an alternative terminology for the understanding of individual and society, but rather the cultivation and nurture of an epistemology of the social which refuses to create other binary and exclusive oppositions.

Post-structuralism suggests that we need to focus on the relation between dichotomies, putting them into quotation marks. It seemed to me a necessary step in the process of transition from a dualist thought to a monist understanding of social reality. Post-structuralism leaves the dichotomies under erasure.[3] Furthermore, post-structuralist anthropological theories slit open the dichotomies to show the conceptual logics which construct the relation between oppositional pairs in modernist Euro-American thought. To my eye, post-structuralism can also operate as a kind of epoché, where the metaphors and analogies of social and anthropological theory are minutely unpacked, opened and, therefore, denaturalised.

I understand that such a procedure could be accurately characterised as phenomenological because it opens a phenomenological fissure in structuralism, challenging the ontological assumption of the oppositional pairs as disembodied universals. Hence, there is a possible dialogue between a post-structural approach and a phenomenology of social life. Nevertheless, such an operation became only a path through which I passed in order to avoid dualisms. It was a necessary step in the labyrinth I crossed. Therefore, I do not want to linger there, but rather keep walking towards a monist understanding of the social worlds, walking as long as it takes. And I will walk through different paths, even if they look like labyrinths because their end is unknown.

The monist notion of the social world seems to be relevant beyond the ethnographic material from which it emerges. It can become more than one methodological tool for analysing the specific idea that self-transformation is the transformation of the world. To describe the resonance of the person with the experienced world, I introduce the notion of social membrane, understood as a point of convergence between the notion of flesh, the Husserlian notion of lifeworld and the notion of social world. This social membrane is intimately entwined with what I have called lived social world. The lived social world is the social membrane – this interstitial space that entangles person and environment – which is transformed and moved as people transform and move. The people with whom I dialogued conceive and live such an experience. But this intricate entanglement of person and world overcomes every conception that externalises the social as a realm transcendent to the set of cultural worlds in which people live.

Each being is more than a line. Each being is a thread of lifelines entangled in social worlds, which are, in turn, a meshwork in which the vibration of one's movements expands and contracts the movement of other. Like the waves generated by stones thrown in the river they annul or increase. Or like the sound vibrations of voices and musical instruments emitted in the same pitch, which meet the same resonant tone or strengthen themselves mutually in the same direction. Maybe, this is how lived social worlds are transformed.

Notes

1 Regarding the fractal social relations, the anthropologist Marylin Strathern, using the reticular vocabulary, would say that "person-to-person networks that succeed by replicating the condition under which persons relate to one another, work, as relations do, holographically (Strathern 1995: 29). The *garia* notion of the person postulates that "the person can be conceptualised as a vessel, like a canoe containing matrikin, adorned on the outside of its relations with others" (Strathern 1992: 98). The social is immanent to the person, and it is a transcendental ontological substrate in relation to the individual. The immanence of the social in the person characterises a system too much distinct from Euro-American holism and individualism, such as analysed by Dumont (1986).

2 As Little (2006: 96) once suggested, one of the challenges of research in the field of political ecology is exactly the search for transversal connections and interrelations through the "fractal analogy", which enables overcoming "systemic approaches in which every level is hierarchically and functionally fit with the other" and also "neo-Marxist approaches in which the superior levels control and determine what happens on the inferior levels". Thus, it is possible to "account in a *sui generis* manner the contingent factors that combine with structuring factors".

3 I thank Professor Luis Felipe Kojima Hirano who, drawing inspiration from the article of Maluf (2013), shared with me the idea of "erasure" as a methodological key in post-structuralist theories. Hirano understands that it is necessary to put concepts under erasure, especially "the dualities which were once universalised" (personal communication, 25/06/2021).

References

Asad, T. 1993. *Genealogies of religion: Discipline and reasons of power in Christianity and Islam*. Baltimore e Londres: The Johns Hopkins University Press.

Dumont, L. 1986. *Essays on individualism: Modern ideology in anthropological perspective*. Chicago: University of Chicago Press.

Gonçalves Brito, L. 2020. "Spirituality and ecology within the phenomenological realm of secularity". *Revista del CESLA* 26: 307–326.

Little, P. 2006. "Ecologia política como etnografia: Um guia teórico". *Horizontes Antropológicos* 12(25): 85–103.

Maluf, S. W. 2013. "Por uma antropologia do sujeito: Da pessoa aos modos de subjetivação". *Campos-Revista de Antropologia* 14(1/2): 131–158.

Steil, C. A., and Carvalho, I. C. M. 2021. "Na 'carne do mundo': Imanência, subjetivação e espiritualidades ecológicas". *Lusotopie* XX(1–2): 1–15.

Strathern, M. 1992. *Reproducing the future: Essays on anthropology, kinship and the new reproductive technologies*. Manchester: Manchester University Press.

——— 1995. *The relation: Issues in complexity and scale*. Cambridge, MA: Prickly Pear Press.

Veer, P. 2009. "Spirituality in modern society". *Social Research* 76(4): 1097–1120.

Index

Note: *Italic* page numbers refer to figures and page numbers followed by "n" denote endnotes.

For Product Safety Concerns and Information please contact our EU
representative GPSR@taylorandfrancis.com
Taylor & Francis Verlag GmbH, Kaufingerstraße 24, 80331 München, Germany

www.ingramcontent.com/pod-product-compliance
Ingram Content Group UK Ltd.
Pitfield, Milton Keynes, MK11 3LW, UK
UKHW020113240926
475569UK00023BA/8
9781032458205